INTRODUCTION TO
PIPE STRESS
ANALYSIS

INTRODUCTION TO
PIPE STRESS ANALYSIS

Sam Kannappan, P.E.
Engineer
Tennessee Valley Authority
Knoxville, Tennessee

A WILEY-INTERSCIENCE PUBLICATION
JOHN WILEY & SONS
New York • Chichester • Brisbane • Toronto • Singapore

Library of Congress Cataloging in Publication Data:

Kannappan, Sam.
  Introduction to pipe stress analysis.

  "A Wiley-interscience publication."
  Includes index.
  1. Pipe lines—Design and construction.
2. Strains and stresses. I. Title.

TJ930.K323 1986   621.8′672   85-17873
ISBN 0-471-81589-6

Printed in the United States of America

10 9 8 7 6 5 4 3 2 1

PREFACE

Until 1967 piping design was performed primarily using rule-of-thumb layout design procedures and preanalyzed piping layout data in tabular form. The publication of ANSI B31.1-1967 Power Piping Code and the availability of analysis computer programs have introduced cost-effective piping design.

The objective of this book is to present a practical approach to analytical piping design. It is intended to be used by engineers in the industry and students interested in piping design. Knowledge of applied mechanics and strength of materials is a must for understanding this book.

The text contains many illustrations, code equations, tables, and examples. Worked out example problems are included to assist the reader in understanding the principles discussed in each chapter. Exercises and references are given at the end of each chapter.

Piping analysis topics, such as support stiffness, overlapping, decoupling of branch lines, wind loads, and other advanced topics, are covered in another book entitled *Advanced Pipe Stress Analysis* by the same author and publisher.

I am indebted to many organizations, including the American Society of Mechanical Engineers and the Expansion Joint Manufactures' Association, for granting permission to reproduce design, tables, and graphs. I thank all my friends and the members of my own family, my wife Meena, sons Ramesh, Narayanan, Ram, daughter Abirami, and my brother S. Narayanan, for their support for me in writing this book.

Sam Kannappan

Knoxville, Tennessee
December 1985

CONTENTS

Contents

CHAPTER ONE

PIPE STRESS ANALYSIS

Pipe stress analysis provides the necessary technique for engineers to design piping systems without overstressing and overloading the piping components and connected equipment. The following terms from applied mechanics are briefly discussed (not defined) here to familiarize the engineer with them.

FORCES AND MOMENTS ON A PIPING SYSTEM

FORCE: The force is a vector quantity with the direction and magnitude of the push (compression), pull (tension), or shear effects.

MOMENT: Moment is a vector quantity with the direction and magnitude of twisting and bending effects.

Forces and moments acting on the piping system due to different types of loadings, such as thermal expansion and dead weight, will be discussed later in detail.

Stress is the force per unit area. This change in length divided by the original length is called strain.

Stress–Strain Curve for Ductile and Nonductile Material

For a ductile material, such as ASTM A53 Grade B, the stress–strain curve is given in Figure 1.1. Until the proportional limit is reached, variation of stress in the material with respect to strain follows a straight line. Hooke's law defines the slope as Young's modulus of elasticity E. Ultimate tensile stress is the highest stress the material can withstand. Yield strength is the point on

1

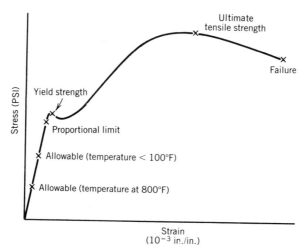

FIGURE 1.1 Typical stress–strain curve for ductile material (ASTM A53 Grade B).

the curve at which any further strain will cause permanent deformations to stressed elements. Allowable stress is the yield strength divided by factor of safety.

A typical stress–strain curve for a nonductile material like cast iron is given in Figure 1.2. The stress–strain diagram for a given piping material shows the limitations on stress to avoid permanent deformation or rupture.

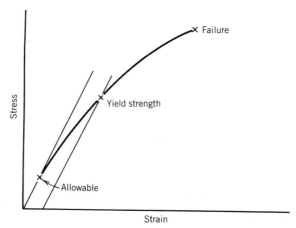

FIGURE 1.2 Typical stress–strain curve for nonductile material (cast iron).

Common Piping Materials

A list of common piping materials under severe cyclic conditions is given next (reference 1):

Pipe for Severe Cyclic Conditions

Only the following pipe* shall be used under severe cyclic conditions:

(a) *Carbon Steel Pipe*

API 5L, Seamless
API 5L, SAW, Factor (E) 0.95 or greater
API 5LX 42, Seamless
API 5LX 46, Seamless
API 5LX 52, Seamless
ASTM A53, Seamless
ASTM A106
ASTM A333, Seamless
ASTM A369
ASTM A381, Factor (E) 0.90 or greater
ASTM A524
ASTM A671, Factor (E) 0.90 or greater
ASTM A672, Factor (E) 0.90 or greater
ASTM A691, Factor (E) 0.90 or greater

(b) *Low and Intermediate Alloy Steel Pipe*

ASTM A333, Seamless
ASTM A335
ASTM A369
ASTM A426, Factor (E) 0.90 or greater
ASTM A671, Factor (E) 0.90 or greater
ASTM A672, Factor (E) 0.90 or greater
ASTM A691, Factor (E) 0.90 or greater

(c) *Stainless Steel Alloy Pipe*

ASTM A268, Seamless
ASTM A312, Seamless

* From ANSI/ASME B31.3, Section 305.23, 1980 edition.

ASTM A358, Factor (*E*) 0.90 or greater
ASTM A376
ASTM A430
ASTM A451, Factor (*E*) 0.90 or greater

(d)   *Copper and Copper Alloy Pipe*

ASTM B42
ASTM B466

(e)   *Nickel and Nickel Alloy Pipe*

ASTM B161
ASTM B165
ASTM B167
ASTM B407

(f)   *Aluminum Alloy Pipe*

ASTM B210, Tempers 0 and H112
ASTM B214, Tempers 0 and H112

For mechanical properties and chemical composition of each one of the above materials, see ASTM standards (reference 2).

Special piping materials include inconel, hastelloy, zirconium, and aluminum alloys. Selection of a specific material depends upon the process temperature and its corrosion properties. Sizing of the piping depends upon volume flow with minimum flow friction (reference 8).

STATIC AND DYNAMIC LOADS

Loadings affecting the piping system can be classified as primary and secondary. Primary loading occurs from sustained loads like dead weight. Primary loads are called non-self limiting loads. An example of a secondary loading (self limiting) is a thermal expansion load. Because different piping codes define the piping qualification criteria in slightly different way, each code will be addressed separately later.

Static loadings include:

1.   Weight effect (live loads and dead loads).
2.   Thermal expansion and contraction effects.
3.   Effects of support, anchor, and terminal movements.
4.   Internal or external pressure loading.

Live loads under weight effect include weight of content, snow, and ice loads. Dead loads consist of weight of piping valves, flanges, insulation, and other superimposed permanent loads.

Dynamic loadings include:

1. Impact forces.
2. Wind.
3. Seismic loads (earthquake).
4. Vibration.
5. Discharge loads.

Piping Material Properties

Thermal effects include thermal loads that arise when free thermal expansion or contraction is prevented by supports or anchors, loads due to temperature gradients in thick pipe walls, and loads due to difference in thermal coefficients of materials as in jacketed piping. The *coefficient of linear expansion* of a solid is defined as the increment of length in a unit length for a change in temperature of one degree. The unit is microinches per inch per °F. The unit for the mean coefficient of thermal expansion between 70°F (installation temperature) and the given temperature is given as inches of expansion per 100 ft of pipe length in Table A1 of Appendix (values are from ASME B31.3 Piping Code). To convert from inch/inch/°F to inch/100 ft, the following relation may be used:

Expansion coefficient (in./100 ft)

$$= (\text{coefficient}) \times 12 \times 100 \; (\text{design temp.} - \text{installation temp.}) \quad (1.1)$$

Young's modulus or modulus of elasticity E is unit stress divided by unit strain. For most structural materials the modulus of elasticity for compression is the same as for tension. Value of E decreases with an increase in temperature. Table A2 of Appendix gives E values for piping materials for the normal temperature range. The ratio of unit lateral contraction to unit axial elongation is called *Poisson's ratio*. Codes allow a value of 0.3 to be used at all temperatures for all metals.

SPECIFIC GRAVITY: The specific gravity of a solid or liquid is the ratio of the mass of an equal volume of water at some standard temperature (physicists use 39°F and engineers use 60°F). The specific gravity of gases is usually expressed in terms of hydrogen and air; it is a number without a unit.

DENSITY: The density ρ is the mass per unit volume of the fluid. The unit is lb/in.^{3} For example, density of carbon steel is 0.283 lb/in.^{3} See Table 1.1.

TABLE 1.1 Poisson's Ratio and Density of Piping Materials

Material Type	Density (lb/in.^{3})	Poisson's Ratio
Carbon steel with 0.3% carbon or less	0.283	0.288
Austenitic steels (SS)	0.288	0.292
Intermediate alloy steel 5% Cr Mo–9% Cr Mo	0.283	0.292
Brass (66% Cu–34% Zn)	0.316	0.331
Aluminum alloys	0.100	0.334

SPECIFIC WEIGHT: The specific weight ω is the weight per unit volume. The interrelation of density and specific weight is $\omega = g\rho$, where g is acceleration due to gravity.

Table 1.1 gives values of Poisson's ratio and density for common piping material.

Example

1. Find the linear thermal expansion (in./100 ft) between 70 and 392°F for carbon steel. Coefficient for 375°F = 2.48 in./100 ft (values from Appendix Table A1).
   Coefficient for 400°F = 2.70 in./100 ft
   Difference per degree in expansion = (2.7 − 2.48)/25 = 0.0088
   By linear interpolation, expansion for

   $$392°F = 2.48 + (392 − 375)(0.0088)$$
   $$= 2.63 \text{ in./100 ft}$$

2. Find the modulus of elasticity for austenitic steel at (a) −200°F, (b) 70°F, and (c) 625°F.
   E at 200°F = 29.9×10^6 psi (read from Appendix Table A2)
   E at 70°F = 28.3×10^6 psi
   E for 625°F should be interpolated between values of 600°F and 700°F
   E at 600°F = 25.4×10^6
   E for 700°F = 24.8×10^6
   E for 625°F is $25.4 − 25((25.4 − 24.8)/100) = 25.4 − 0.15 = 25.25 \times 10^6$ psi

Note that the E value decreases with increase in temperature. Lower values of Young's modulus means that the flexibility is higher. Use of hot modulus E_h is permitted in calculating forces and moments at the equipment nozzles. However, the higher value (at 70°F or at installation temperature) should be used in stress calculations.

PIPING SPECIFICATION

Piping specification is written for each service such as steam, air, oxygen, and caustic. The specification contains information about piping material, thickness, recommended valves, flanges, branch connection, and instrument connection. Figure 1.3 shows a specification for caustic service.

Example

An 8 in. pipe needs a pipe with thickness of 80 schedule (which allows for $\frac{1}{8}$ in. corrosion allowance and maximum internal pressure of 200 psig up to 150°F) with a bevel-edged A53 Grade B seamless. The globe valve used is crane $351\frac{1}{4}$ (reference 1 in Chapter 9). The flanges are of 150 psi pressure rating with raised face and weld neck slip on type. The material of the flange is A-105 (per standard ANSI B16.5). The requirement for the branch connection (here weldolet or tee) is given on the branch connection table. For an 8 in. header and a 3 in. branch, the weldolet is required for given internal pressure. The pressure and temperature conditions in the pipeline should always be within (inside the hatched line) the pressure–temperature curve given in the specification.

Flexibility

Piping systems should have sufficient flexibility so that thermal expansion or contraction or movements of supports and terminal points will not cause:

1. Failure of piping or support from overstress or fatigue.
2. Leakage at joints.
3. Detrimental stresses or distortion in piping or in connected equipment (pumps, vessels, or valves, for example) resulting from excessive thrusts or moments in the piping.

Flexibility denotes the measurement of the presence of necessary piping length in the proper direction. The purpose of piping flexibility analysis is to produce a piping layout that causes neither excessive stresses nor excessive end reactions. To achieve this, layout should not be stiff. It is also not desirable to make the system unnecessarily flexible because this requires excess materials, thus increasing initial cost. More length with many bends increases pressure drop, which increases operating cost.

Size inch	Piping	Gate ball plug	Globe	Check	Mechanical joints	Fittings	Reduced	Full size	Size
1/2	Sch 160	V-BOCB 125 psi	V-IGHT 150 psi	V-9CNY 1000 psi	300 psi screwed	300 psi	Reducing	Straight	1/2
3/4	ASTM A-106 GR B	Screwed all iron	Screwed all iron	MI crane 346 1/2	Gasket union steel	MI screwed	screwed tee	screwed	3/4
1	Seamless	Crane 484 1/2	Crane 355 1/2		seats			tee	1
1 1/2									1 1/2
2	Sch 80	V BOCC 125 psi	V-BGHU 125 psi	V-BCHZ 125 psi	150 psi raised face	Sch 80 butt weld	1 1/2 in. and	as shown in	2
3	with bevel	FF all iron	FF all iron	FF all iron	Flange ASTM	ASTM A-234	smaller	table below	3
4	edged A-53 GR B	Crane 475 1/2	Crane 351 1/4	Crane 373 1/2	A 105 weld neck	Seamless	3000 psi threadolet		4
6	Seamless				slip on except		2 in. and larger		6
8					at fittings		see table		8
10					see note 1		below		10

Valves   **Branch Connections**

Orifice assembly

1/2 " V-BOCB

Instr. piping

1/2 in. × 3 "TBE

1/2 × 7" TBE

Flow

Piping

Instr.

300 psi raised face weld neck orifice flange with screwed taps
Lines 2 in. and larger

Vent and drains

3/4 in. bar stock Plug

3/4 in. TOL

3/4" × 3 in. TRE (Typ-2)

3/4 in. V-BOCB

3/4 in. SCRD cap

Bolting:
A 193 GR B-7 alloy steel stud bolts with 2A-194 GR2 -H heavy hex. nuts. Note 1
Gaskets: 1/16 in. asbestos full face

Pressure connections

3/4 in. V-BOCB

PG

3/4 in. Tol

Piping

Inst.

Instrumentation notes:
1. Instrumentation connections are typical

Temperature connections

TW

1 in. TOL.

Instr. piping

Lines 3 in. and larger

TW

1 in. EOL

FLow

Instr. piping

Lines 2 in. and larger.
Increase smaller lines

8

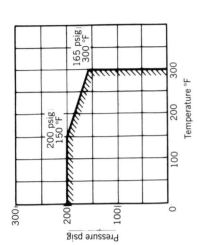

Code B 31.3

Corr. allow $\frac{1}{8}$ in.

| Rating 125 psi carbon steel | Service Caustic, Sludge Slag, Quench water | Piping Spec. No. |

FIGURE 1.3 Typical piping specification.

Notes:
1. Bolt lengths and flanges per ANSI B16.5 std.
2. Use teflon tape for screwed connections
3. Post heat treat all welds
4. No brass allowed in quench water line

Branch connection table:
W weldlolet   T tee

HEADER

2	3	4	6	8	10	
T						
W	T					
W	W	T				
W	W	W	T			
W	W	W	W	T		
W	W	W	W	W	T	
2	3	4	6	8	10	

BRANCH

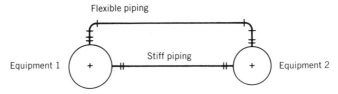

FIGURE 1.4 Flexible and stiff piping.

Figure 1.4 shows examples of stiff and flexible piping. When a piping is subjected to change in temperature and if the pipe is not restrained from expansion, no stresses are developed and the pipe just expands or contracts. When the pipe is restrained, stresses and forces of considerable magnitude are created. For example, at a refinery near Houston, Texas, when two axial restraints were present in a straight steam line (see Fig. 1.13), the bending of a large support frame and the failure of a pipe at the shoe-pipe weld area occurred.

The thermal force that is developed when both ends of a hot piping are restrained is enormous and is also independent of the length of piping.

$$\text{Thermal force} = E(\text{strain due to expansion})(\text{metal area}) \qquad (1.2)$$

Example

Calculate the force developed in a 10 in. sch 40 carbon steel pipe A53 Grade B subjected to 200°F from an installation temperature of 70°F.

The metal area of a 10 in. sch 40 pipe is 11.9 sq in. (Appendix Table A4). The expansion coefficient at 200°F is 0.99 in./100 ft (Appendix Table A1).

$E = 27.9 \times 10^6 \, \text{psi}$ (Appendix Table A2)

$$F = E\alpha A = 27.9 \times 10^6 \times \frac{0.99}{100 \times 12} \times 11.9 \quad \left[\text{units: } \frac{\text{lb}}{\text{in.}^2} \left(\frac{\text{in.}}{\text{in.}} \right) \text{in.}^2 = \text{lb} \right]$$

$$= 273{,}908 \, \text{lb}$$

The layout of a piping system provides inherent flexibility through changes in direction. The stiff piping system shown in Figure 1.4 can be made flexible in different ways. Figure 1.5 shows the inclusion of an expansion loop if space permits. An expansion joint (Fig. 1.6) may be added (see Eq. 5.4 for

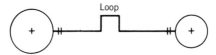

FIGURE 1.5 Piping with expansion loop.

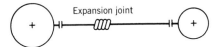

FIGURE 1.6 Piping with expansion joint.

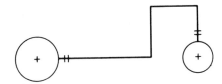

FIGURE 1.7 Leg provided by turning equipment.

pressure thrust calculation) or the equipment may be turned by 90 degrees and thus provides the leg to absorb the expansion, as shown in Figure 1.7.

When a piping system lacks built-in changes in direction, the engineer should consider adding flexibility by one or more of the following means: bends, loops or offsets, swivel joints, corrugated pipe, expansion joints of the bellows or slip joint type, or other devices permitting angular, rotational, or axial movements. Expansion joints and expansion loops will be discussed in detail in Chapter 5.

EXPLANATION OF TERMS RELATED TO PIPE SUPPORTS

ANCHOR: A rigid restraint providing substantially full fixity for three translations and rotations about the three reference axes. A large number in the order of 10^{12} lb/in. is assumed for translational stiffness in the digital computer programs to simulate the fixity. The details of a structural anchor may be obtained from each company's pipe support standard.

BRACE: A device primarily intended to resist displacement of the piping due to the action of any forces other than those due to thermal expansion or to gravity. Note that with this definition, a damping device is classified as a kind of brace.

CONSTANT-EFFORT SUPPORT: A support capable of applying a relatively constant force at any displacement within its useful operating range (e.g., counterweight or compensating spring device).

DAMPING DEVICE: A dashpot or other frictional device that increases the damping of a system, offering high resistance against rapid displacements caused by dynamic loads while permitting essentially free movement under very gradually applied displacements (e.g., snubber).

HANGER: A support by which piping is suspended from a structure, and so on, and which functions by carrying the piping load in tension.

LIMIT STOP:　A device that restricts translatory movement to a limited amount in one direction along any single axis. Paralleling the various stops there may also be double-acting limit stops, two-axis limit stops, and so on.

RESILIENT SUPPORT:　A support that includes one or more largely elastic members (e.g., spring).

RESTING OR SLIDING SUPPORT:　A device providing support from beneath the piping but offering no resistance other than frictional to horizontal motion.

RESTRAINT:　Any device that prevents, resists, or limits the free movement of the piping.

RIGID (SOLID) SUPPORT:　A support providing stiffness in at least one direction, which is comparable to that of the pipe.

STOP:　A device that permits rotation but prevents translatory movement in at least one direction along any desired axis. If translation is prevented in both directions along the same axis, the term double-acting stop is preferably applied. Stop is also known as "Bumper."

SUPPORT:　A device used specifically to sustain a portion of weight of the piping system plus any superimposed vertical loadings.

TWO-AXIS STOP:　A device which prevents translatory movement in one direction along each of two axes.

Once a complete (weight, thermal plus pressure, and thermal plus pressure plus weight) analysis of the piping system has been conducted, support modifications can be made very easily.

When a pipe line moves as a result of thermal expansion, it is necessary that flexible hangers be provided that support the piping system throughout its thermal cycle. Three types of hangers are generally employed:

1.　Rigid support or rod hangers that supposedly prevent any movement along the axis of the hanger. Rod hangers are used when the free thermal deflections are small enough so that their restraint of movement does not produce excessive reactions in the piping system.

2.　Variable support or spring hangers provide a supporting force equal to hot load (reference 6) while allowing deflection.

3.　Constant support or constant effort hangers that provide an essentially constant supporting force throughout the thermal cycle. Ideally, constant support hangers do not restrain the free movement of the system and therefore do not increase the piping stresses.

THE GUIDED CANTILEVER METHOD

One of the simplified methods used in piping design is known as the guided cantilever method, because deflections are assumed to occur in a single-

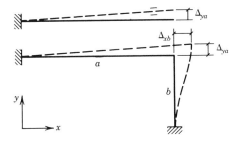

FIGURE 1.8 Guided cantilever approximation.

plane system under the guided cantilever approximation, as shown in Figure 1.8. The deflection capacity of a cantilever under this assumption can be given by Eq. 1.3 (reference 3):

$$\Delta = \frac{144 L^2 S_A}{3 E D_o} \tag{1.3}$$

where Δ = permissible deflection, inches
S_A = allowable stress range, psi (given by Eq. 4.1)
L = length of leg needed to absorb the expansion, feet
D_o = outside diameter of pipe, inches.

The limitations of the guided cantilever method are:

1. The system has only two terminal points and it is composed of straight legs of a pipe with uniform size and thickness and square corner intersections.
2. All legs are parallel to the coordinate axes.
3. Thermal expansion is absorbed only by legs in a perpendicular direction.
4. The amount of thermal expansion that a given leg can absorb is inversely proportional to its stiffness. Because the legs are of identical cross section, their stiffness will vary according to the inverse value of the cube of their lengths.
5. In accommodating thermal expansion, the legs act as guided cantilevers, that is, they are subjected to bending under end displacements; however, no end rotation is permitted, as shown in Figure 1.8.

As a further refinement of this method, a correction factor that allows for reducing the bending moment, due to the rotation of the leg adjacent to the one considered, can be used (reference 3).

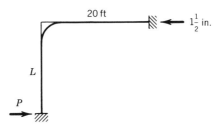

FIGURE 1.9   Anchor with initial movement.

Example

Calculate leg L required for the two anchor problem and force P given in Figure 1.9.

Pipe outside diameter $= 4\frac{1}{2}$ in.; thickness $= 0.237$ in.

Expansion coefficient $= 4$ in./100 ft

Stress range $= S_A = 15,000$ psi

Cold modulus $= 27.9 \times 10^6$ psi

Deflection $\Delta = 1\frac{1}{2} + 20(4/100) = 2.3$ in.

Rearranging Eq. 1.3 (guided cantilever method):

$$L = \sqrt{\frac{3ED_o\,\Delta}{144S_A}} = \sqrt{\frac{3 \times 27.9 \times 10^6 \times 4.5 \times 2.3}{144(15,000)}} = 20.03 \text{ ft}$$

$$\text{Bending stress} = S_b = \frac{\text{moment}}{Z} = \frac{PL}{2Z}$$

$$\text{Mean radius } r = \frac{1}{2}\left[\frac{4.5 + 4.5 - 2(0.237)}{2}\right] = 2.13 \text{ in.}$$

$$Z = \text{section modulus} = \pi r^2(\text{thickness}) = \pi(2.13)^2(0.237) = 3.38 \text{ in.}^3$$

$$\text{Force } P = \frac{2S_bZ}{L} = \frac{2(15,000)(3.38)}{20.03(12)} = 421.8 \text{ lb}$$

COMPARISON OF SIMPLIFIED ANALYSIS METHODS

Results obtained from other simplified methods and the digital computer aided piping analysis are compared here. However, each method is not fully explained because the references give a detailed explanation and they also need charts and graphs for their solution.

To understand the differences between each of the methods, results for three problems (Table 1.3) for range of diameters 6–24 in. are presented here (reference 4).

Methods

1. Tube turns (reference 5)
2. ITT Grinnell (reference 6)
3. M. W. Kellogg (reference 3)
4. Digital computer solution including bend flexibility factors (reference 7)
5. Digital computer solution using square corner approach (not including the bend flexibility)

Table 1.2 includes the range of diameters (6–24 in.), wall thickness, and moment of inertia I used in the calculations. Table 1.3 shows the configuration of a U loop (expansion loop) an L shape, and a Z shape. The maximum bending stress is also given for each method.

Figure 1.10 shows the variation of bending stress with area moment of inertia I for the loop. Here I was selected instead of diameter because I also includes the effect of wall thickness. As can be seen the Grinnell method gives very highly conservative results. Expansion loops are further discussed in Chapter 5.

Figure 1.11 shows the variation of bending stress for the L shape. The Kellogg method gives higher stress values. Figure 1.12 demonstrates the variation of bending stress with moment of inertia for the Z shape. The digital computer solution using EZFLEX computer program gives lower numbers, which is understandable because the other methods are meant to be conservative. The Kellog method is discussed in detail in Chapter 5 (Eqs. 5.2 and 5.3).

TABLE 1.2   Pipe Sizes Used in Comparison of Simplified Methods

Pipe O.D. (in.)	Sch	Inside Diameter	Wall Thickness	Moment of Inertia I (in.^{4})	Modulus of Section Z, (in.^{3})
6.625	40	6.025	0.280	28.14	8.50
8.625	40	7.981	0.332	72.50	16.81
10.75	20	10.250	0.250	113.70	21.16
12.75	Std.	12.000	0.375	279.30	43.80
14.00	20	13.376	0.316	314.30	44.90
16.00	Std.	15.250	0.375	562.10	70.30
18.00	20	17.376	0.312	678.00	75.51
20.00	Std.	19.250	0.375	1114.00	111.4
24.00	Std.	23.25	0.375	1943.0	161.9

TABLE 1.3 Comparison of Maximum Bending Stress from Different Methods, psi

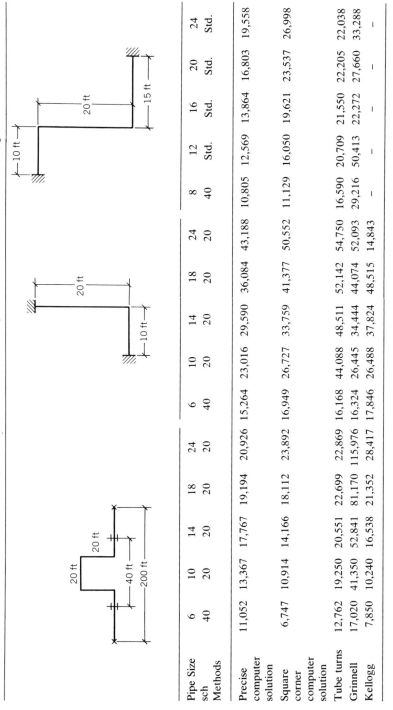

Pipe Size sch Methods	6 40	10 20	14 20	18 20	24 20	6 40	10 20	14 20	18 20	24 20	8 40	12 Std.	16 Std.	20 Std.	24 Std.
Precise computer solution	11,052	13,367	17,767	19,194	20,926	15,264	23,016	29,590	36,084	43,188	10,805	12,569	13,864	16,803	19,558
Square corner computer solution	6,747	10,914	14,166	18,112	23,892	16,949	26,727	33,759	41,377	50,552	11,129	16,050	19,621	23,537	26,998
Tube turns	12,762	19,250	20,551	22,699	22,869	16,168	44,088	48,511	52,142	54,750	16,590	20,709	21,550	22,205	22,038
Grinnell	17,020	41,350	52,841	81,170	115,976	16,324	26,445	34,444	44,074	52,093	29,216	50,413	22,272	27,660	33,288
Kellogg	7,850	10,240	16,538	21,352	28,417	17,846	26,488	37,824	48,515	14,843	–	–	–	–	–

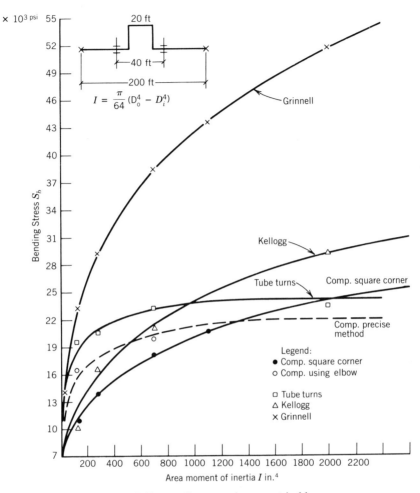

FIGURE 1.10 Bending stress in symmetrical loop.

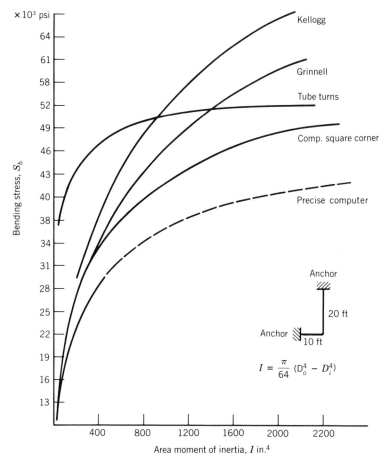

FIGURE 1.11. Bending stress in L-shaped piping.

18

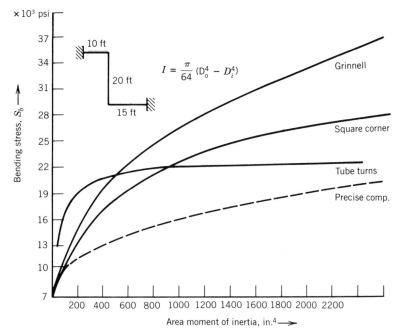

FIGURE 1.12   Bending stress in Z-shaped piping.

EXERCISES

1. (a) Find total expansion for intermediate alloy steel (5Cr Mo through 9 Cr Mo) pipe at temperatures of (1) −55°F, (2) 431°F, (3) 1572°F. If the temperature given is out of range for the material, suggest suitable material for that temperature. Consider length of 120 ft.
    (b) Find for austenitic steel the following at installation temperature:
    (1)   Young's modulus
    (2)   Poisson's ratio
    (3)   Density.
    (c)   Calculate total elongation in 132 ft of pipe made of carbon steel subjected to 645°F.

2. (a)   Find E values for low chrome steel at −115°F, 70°F, and 800°F. Explain the effect of temperature on E value.
    (b)   Find cold and hot stresses for ASTM A53 Grade B pipe at 70°F and 625°F.

3. Calculate the thermal force developed in the piping that is fixed at both ends as shown in Figure 1.13. It consists of an 8 in. sch 40, carbon steel pipe with operating temperature 300°F. Use Eq. 1.2.

    α = coefficient of thermal expansion at 320°F = 1.82 in./100 ft

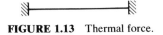

FIGURE 1.13   Thermal force.

FIGURE 1.14   Unequal legs piping with L-shape.

4. Calculate the stress of the layout in Figure 1.14. It consists of a 10 in. sch 40, carbon steel pipe of A53 Grade B material at 500°F.

$$S_c = 20,000 \, \text{psi} \qquad S_h = 17,250 \, \text{psi}$$

5. A 10 in. sch 40 carbon steel pipe with A53 Grade B material has a temperature of 200°F. The allowable stress $S_c = S_h = 20,000$ psi. Calculate leg L needed in Figure 1.15.

6. Two equipment nozzles have thermal movement and layout as shown in Figure 1.16. What will be the length L?

   The carbon steel pipe has a nominal diameter of 8 in. and $\alpha = 1.82$ in./100 ft.

$$S_A = 18,000 \, \text{psi} \qquad E = 27.9 \times 10^6 \, \text{psi}$$

7. Two vessels are connected by piping as shown in Figure 1.17. What is the length required for the leg? What is the force and moment?

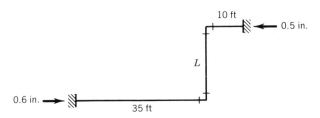

FIGURE 1.15   A Z-shaped piping with initial anchor movements.

FIGURE 1.16   Determination of leg required.

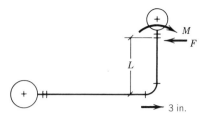

FIGURE 1.17 Calculation of force and moment at anchor.

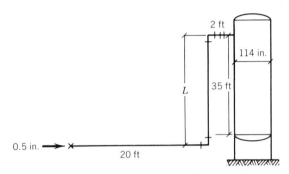

FIGURE 1.18 Piping connected to a vessel.

For a 6 in. sch 40 carbon steel pipe A53 Grade B, the linear expansion is 3 in. Allowable stress range $S_A = 28,000$ psi.

8. A vessel has an average operating temperature of 500°F. With a line from the vessel nozzle going to an equipment as shown in Figure 1.18, what should be the length L?

It is a 12 in. sch 40 pipe with a temperature of 400°F. The pipe is of A53 Grade B material. $S_c = 20,000$ psi and $S_h = 16,350$ psi. (In practical cases, L is limited by tower height.)

REFERENCES

1. ANSI/ASME B31.3-1980 *Chemical Plant and Petroleum Refinery Piping*.
2. ASTM Annual Book of ASTM Standards, *Different Parts for Different Materials*.
3. M. W. Kellogg, *Design of Piping Systems*. New York:
4. Estrems, Fernando and S. Kannappan, "Comparison of results from different simplified methods with digital computer calculations."
5. Tube Turns Division of Chemetron Corp. "Piping Engineering, Line Expansion and Flexibility."
6. ITT Grinnell Industrial Piping. "Piping Design and Engineering."
7. EZFLEX Piping Flexibility Analysis Program.
8. Crane Company. "Flow of Fluids."

CHAPTER TWO

DESIGN OF PRESSURE COMPONENTS

CALCULATION OF MINIMUM WALL THICKNESS OF A PIPE

Piping codes require that the minimum thickness t_m, including the allowance for mechanical strength, shall not be less than the thickness calculated using Eq. 2.1.

$$t_m = \frac{PD_o}{2(SE_q + PY)} + A$$

$$= t + A \qquad\qquad (2.1)$$

where t_m = minimum required wall thickness, inches
$\quad t$ = pressure design thickness, inches
$\quad P$ = internal pressure, psig
$\quad D_o$ = outside diameter of pipe, inches
$\quad S$ = allowable stress at design temperature (known as hot stress), psi (see Appendix Table A3)
$\quad A$ = allowance, additional thickness to provide for material removed in threading, corrosion, or erosion allowance; manufacturing tolerance (MT) should also be considered.
$\quad Y$ = coefficient that takes material properties and design temperature into account. For $t < d/6$, values of Y are given in Table 2.1. For temperature below 900°F, 0.4 may be assumed.

$$Y = \frac{d}{d + D_o} \qquad \text{if } t \geq \frac{d}{6} \qquad (2.2a)$$

where d = inside diameter = $D_o - 2t$

E_q = quality factor that is the product of casting quality factor E_c, joint quality factor E_j, and structural grade quality factor E_s when applies. Values of E_c range from 0.85 to 1.0 and depends upon the method used to examine the casting quality (see Table 2.2a). Value of E_j ranges from 0.6 to 1.0 (given in Table 2.2b) and depends upon type of weld joint. Values of E_s may be assumed as 0.92.

$$E_q = E_c E_j E_s \qquad\qquad (2.2b)$$

TABLE 2.1   Values of Y Coefficient to Be Used in Eq. 2.1[a]

Temperature (°F) Material	900°F and below	950	1000	1050	1150	1150 and above
Ferritic steels	0.4	0.5	0.7	0.7	0.7	0.7
Austenitic steel	0.4	0.4	0.4	0.4	0.5	0.7
Cast iron	0.4	–	–	–	–	–
Nonferrous metals	0.4	–	–	–	–	–

[a]Reference ANSI/ASME B31.3, Table 304.1.1.

TABLE 2.2a   Increased Casting Quality Factor E_c[a]

Type of Supplementary Examination	E_c
Surface examination (1)	0.85
Magnetic particle method (2)	0.85
Ultrasonic examination (3)	0.95
Type 1 and 2	0.90
Type 1 and 3	1.00
Type 2 and 3	1.00

[a]Reference ANSI/ASME B31.3, Table 302.3.3c.

TABLE 2.2b   Straight and Spiral Longitudinal Weld Joint Quality Factor $E_j{}^a$

Type of Joint	Examination	E_j
Furnace butt weld	As required by specification	0.60
Electric resistance weld	As required by specification	0.85
Electric fusion weld (single butt weld)	As required by specification	0.80
Electric fusion weld (single butt weld)	Spot radiograph	0.90
Electric fusion weld (single butt weld)	100% radiograph	1.00
Electric fusion weld (double butt weld)	As required by specification	0.85
Electric fusion weld (double butt weld)	Spot radiograph	0.90
Electric fusion weld (double butt weld)	100% radiograph	1.00
By ASTM A211 specification	As required by specification	0.75
Double submerged arc-welded pipe (per API 5L or 5LX)	Radiograph	0.95

^{a}Reference B31.3 ANSI/ASME 302.3.4.

Example

Calculate the minimum permissible wall thickness for a 10 in. nominal diameter pipe under 350 psi and 650°F. Material is ASTM A106 Grade B carbon steel, corrosion allowance is 0.05 in., and mill tolerance (MT) is $12\frac{1}{2}\%$.

$$\text{Thickness } t_m = \frac{PD_o}{2(SE_q + PY)} + A \qquad (2.1)$$

$P = {} = 350\,\text{psig} \qquad D_o = 10.75\,\text{in.} \qquad E_q = 1.0$ for seamless pipe

$S = S_h =$ hot allowable stress (tensile) for A106 Grade B $= 17{,}000\,\text{psi}$ (see Appendix A3)

$Y = 0.4$ (because the temperature is less than 900°F)

$$t_m = \frac{350(10.75)}{2(17{,}000 \times 1 + 350 \times 0.4)} + 0.05 = 0.144\,\text{in.}$$

$$\text{Nominal thickness} = \frac{0.144}{(1-\text{MT})} = \frac{0.144}{(1-0.125)} = 0.1648\,\text{in.}$$

From the manufacturer and pipe section properties information, (see Appendix A4) a 10 in. pipe with sch 20 is selected with nominal wall thickness of 0.25 in. For pipes under external pressure see Eqs. 9.10 through 9.13.

Alternate Equations to Calculate Wall Thickness

Looking at Eq. 2.1 again, we see that:

$$t_m = t + A$$

$$= \frac{PD_o}{2(SE_q + PY)} + A \tag{2.1}$$

where t is the pressure design thickness in inches.

Equations 2.3 and 2.4 (Lamé equation) may also be used to calculate t:

$$t = \frac{PD_o}{2SE_q} \tag{2.3}$$

$$t = \frac{D_o}{2}\left(1 - \sqrt{\frac{SE_q - P}{SE_q + P}}\right) \tag{2.4}$$

Equations 2.1, 2.3, and 2.4 are valid for $t < D_o/6$ (thin pipe).

The pipe with $t \geq D/6$ (thick-walled pipe) or $P/SE_q > 0.385$ requires special consideration taking design and material factors into account such as theory of failure, fatigue, and thermal stress (reference 1).

Allowable Working Pressure

The allowable working pressure of a pipe can be determined by Eq. 2.5:

$$P = \frac{2(SE_q)t}{D_o - 2Yt} \tag{2.5}$$

where t = specified wall thickness or actual wall thickness in inches.

For bends the minimum wall thickness after bending should not be less than the minimum required for straight pipe.

Blanks

The pressure design thickness t of permanent blanks is given by the equation:

$$t = d_g \sqrt{\frac{3P}{16SE_q}} + \text{Allowance} \tag{2.6}$$

where d_g = inside diameter of gasket for raised face or flat (plain) face flanges or gasket pitch diameter for ring joint and fully retained gasketed flanges in inches.

Test Pressure

The hydrostatic test pressure at any point in the system should be not less than $1\frac{1}{2}$ times the design pressure. For temperatures above 650°F, the minimum test pressure P_T is given by:

$$P_T = 1.5\left(\frac{S_T}{S}\right)(\text{Design pressure}) \tag{2.7}$$

S_T = allowable stress at 650°F (S_h at 650°F)   (see Appendix Table A3)

S = allowable stress at design temperature   (S_h at design temperature)

Allowable Pressure in Miter Bends

Miter Bends*

An angular offset of 3 degrees or less (angle α in Figure 2.2) does not require design consideration as a miter bend. Acceptable methods for pressure design of multiple and single miter bends are given in (a) and (b) next.

(*a*) *Multiple Miter Bends:* The maximum allowable internal pressure shall be the lesser value calculated from Eqs. 2.8a and b. These equations are not applicable when θ exceeds 22.5 degrees

$$P_m = \frac{SE_q(T-c)}{r_2}\left[\frac{T-c}{(T-c)+0.643 \tan \theta\sqrt{r_2(T-c)}}\right] \tag{2.8a}$$

$$P_m = \frac{SE_q(T-c)}{r_2}\left(\frac{R_1-r_2}{R_1-0.5r_2}\right) \tag{2.8b}$$

(*b*) *Single Miter Bends (or Widely Spaced Miter Bends)*
    (1) The maximum allowable internal pressure for a single miter bend with angle θ not greater than 22.5 degrees shall be calculated by Eq. 2.8a.
    (2) The maximum allowable internal pressure for a single miter bend with angle θ greater than 22.5 degrees shall be calculated by Eq. 2.8c:

$$P_m = \frac{SE_q(T-c)}{r_2}\left[\frac{T-c}{(T-c)+1.25 \tan \theta\sqrt{r_2(T-c)}}\right] \tag{2.8c}$$

*From ASME/ANSI B31.3, Section 304.2.3.

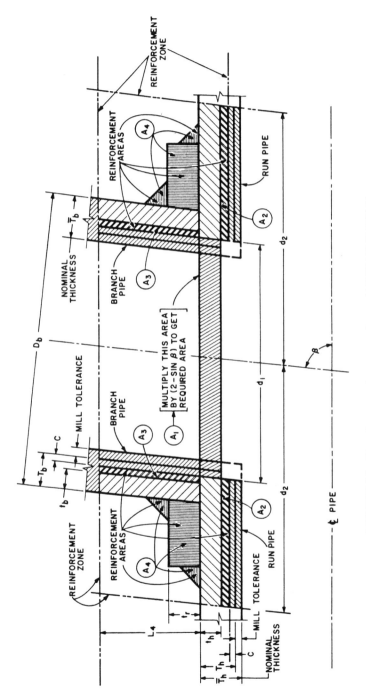

FIGURE 2.1 Branch connection nomenclature (ANSI/ASME B31.3).

27

(3)   The following nomenclature is used in Eqs. 2.8a, 2.8b, and 2.8c for the pressure design of miter bends:

c = corrosion allowance
P_m = maximum allowable internal pressure for miter bends
r_2 = mean radius of pipe using nominal wall T
R_1 = effective radius of miter bend, defined as the shortest distance from the pipe centerline to the intersection of the planes of adjacent miter joints
E_q = quality factor (see Eq. 2.2b)
S = allowable stress at design temperature, psi
T = pipe wall thickness (measured or minimum per purchase specification)
θ = angle of miter cut, degrees
α = angle of change in direction at miter joint = 2θ

For compliance with this code, the value of R_1 shall not be less than that given by Eq. 2.9:

$$R_1 = \frac{A_1}{\tan \theta} + \frac{D_o}{2} \qquad (2.9)$$

where A_1 has the following empirical values (not valid in SI units):

Value of $(T - c)$, in.	Value of A_1
≤ 0.5	1.0
$0.5 < (T - c) < 0.88$	$2(T - c)$
≥ 0.88	$\dfrac{2(T - c)}{3} + 1.17$

See Chapter 4 for further discussion on miter bends.

Example

Calculate maximum allowable internal pressure for the multiple miter bend. Plate thickness is $\frac{1}{2}$ in. Corrosion allowance is zero. Manufacturing allowance is 0.01 in. Miter OD is 36 in.

For two-weld miter (see Fig. 2.2):

$$\theta = \frac{\alpha}{2} = \frac{22\frac{1}{2}}{2} = 11\frac{1}{4}^{\circ}$$

The mean radius of the pipe $= 35.5/2 = 17.75$ in.

The material consists of A312 TP 304 H stainless steel. Temperature is 1310°F.

Allowable hot stress is $S_h = SE_q = 3060$ psi   (From Appendix Table A3).

Interpolate between $S_h = 3700$ psi for 1300°F and $S_h = 2900$ psi for 1350°F. Bend radius, $R_1 = 54$ in. (see Table 4.4).

Using Eq. 2.8a, allowable pressure is

$$P_m = \frac{SE_q(T-c)}{r_2}\left[\frac{T-c}{(T-c)+0.643\tan\theta\sqrt{r_2(T-c)}}\right]$$

$$= \frac{3060(0.5-0.01)}{17.75}\left[\frac{(0.5-0.01)}{(0.5-0.01)+0.643\tan(11.25)\sqrt{17.75(0.5-0.01)}}\right]$$

$$= 58 \text{ psig}$$

Using Eq. 2.8b, the allowable pressure is:

$$P_m = \frac{SE_q(T-c)}{r_2}\left(\frac{R_1-r_2}{R_1-0.5r_2}\right)$$

$$= \frac{3060(0.49)}{17.75}\left(\frac{54-17.75}{54-0.5(17.75)}\right)$$

$$= 67.86 \text{ psig}$$

The maximum allowable pressure for the miter is the smaller of the values calculated above. Thus $P_m = 58$ psig.

REINFORCEMENTS FOR WELDED BRANCH CONNECTIONS*

When a hole is cut in a pipe subjected to internal pressure, the disc of the material that would normally be carrying tensile stresses in the hoop direction is removed and an alternate path must be provided. To achieve this, a simplified "area replacement" or "compensation" approach is used. This method provides for additional reinforcement material, which is within a specified distance from the edge of the hole, equal to the area of the material removed. Reinforcement at branch intersections are also occasionally needed to distribute stresses arising from pipe loads. See discussion of stress intensification factors (SIF) in Chapter 4 for the reduc-

*From ASME/ANSI B31.3. Section 304.3.3.

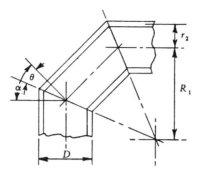

FIGURE 2.2   Nomenclature for miter bends.

tion of the calculated value of SIF when reinforcement was provided. The reinforcement requirement for internal pressure is usually defined in the piping specification of the project. Additional reinforcement may be needed for piping loads.

Figure 2.1 shows pipe run-branch connection (reproduced from B31.3 code). Requirements of the other codes are similar. A number of worked out problems are given in Appendix H of B31.3 code (Reproduced here as Appendix Table A5).

The requirements are not applicable to branch connections in which the smaller angle between branch and run is less than 45 degrees, or in which the axis of the branch does not intersect the axis of the run.

(a) *Nomenclature.* The nomenclature below is used in the pressure design of welded branch connections. It is illustrated in Fig. 2.1, which does not indicate details for construction or welding.

Definitions

b = subscript referring to branch

d_1 = effective length removed from pipe at branch

d_2 = "half width" of reinforcement zone

    = d_1 or $(T_b - c) + (T_h - c) + d_1/2$, whichever is greater, but in any case not more than D_h

h = subscript referring to run or header

L_4 = height of reinforcement zone outside of run pipe

    = $2.5(T_h - c)$ or $2.5(T_b - c) + T_r$, whichever is less

T_r = minimum thickness of reinforcing ring or saddle made from pipe (Use nominal thickness if made from plate.)

    = 0, if there is no reinforcement pad or saddle

t = pressure design thickness of pipe, according to the appropriate wall thickness Eq. 2.1. For welded pipe, when the branch does not intersect the longitudinal weld of the run, the basic allowable stress S for the pipe (see Appendix A3) may be used in determining t_h for the purpose of reinforcement calculation only. When the branch does intersect the longitudinal weld of the run, the allowable stress SE_q of the run pipe shall be used in the calculation. The allowable stress SE_q of the branch shall be used in calculating t_b.

β = smaller angle between axes of branch and run

(b) *Required Reinforcement Area.* The reinforcement area A_1 required for branch connections under internal pressure shall be:

$$A_1 = (t_h d_1)(2 - \sin \beta) \tag{2.10}$$

and under external pressure shall be:

$$A_1 = \frac{(t_h d_1)(2 - \sin \beta)}{2} \tag{2.11}$$

(c) *Reinforcement Area.* The reinforcement area is the sum of areas $A_2 + A_3 + A_4$ defined below, and shall equal or exceed the required reinforcement area A_1.

(1) Area A_2. The area lying within the reinforcement zone resulting from any excess thickness available in the run wall:

$$A_2 = (2 d_2 - d_1)(T_h - t_h - c) \tag{2.12}$$

(2) Area A_3. The area lying within the reinforcement zone resulting from any excess thickness available in the branch pipe wall:

$$A_3 = \frac{2 L_4 (T_b - t_b - c)}{\sin \beta} \tag{2.13}$$

(3) Area A_4. The area of all other metal within the reinforcement zone provided by weld metal and other reinforcement metal properly attached to the run on branch.

Materials used for reinforcement may differ from those of the run pipe, provided they are compatible with the run and branch pipes with respect to weldability, heat treating requirements, galvanic corrosion, thermal expansion, and so on. If the allowable stress for such materials is less than that for the run pipe, the corresponding calculated area must be reduced in the ratio of the allowable stress values before being counted toward the reinforcement area. No additional credit shall be taken for materials having higher allowable stress values than the run pipe.

(d) *Reinforcement Zone.* The reinforcement zone is a parallelogram whose length extends a distance of d_2 on each side of the centerline of the branch pipe and whose width starts at the inside surface of the run pipe (in its corroded condition) and extends to a distance L_4 from the outside surface of the run pipe measured perpendicular to this outside surface.

(e) *Reinforcement of Multiple Openings.* When any two or more adjacent openings are so closely spaced that their reinforcement zones overlap, the two or more openings shall be reinforced* with a combined reinforcement that has an area equal to that required for the separate openings. (See ASME/ANSI, Section 304.3.4 of B31.3 code for reinforcement requirements of extruded outlet headers for further reading.) No portion of the cross section shall be considered as applying to more than one opening, or be evaluated more than once in a combined area. When two or more openings are to be provided with a combined reinforcement, the minimum distance between centers of any two of these openings should preferably be at least $1\frac{1}{2}$ times their average diameter, and the area of reinforcement between them shall be at least equal to 50% of the total required for these two openings. (Pipe Fabrication Institute Standard ES-7 may be consulted for detailed recommendations on spacing between welded nozzles.)

(f) *Rings and Saddles.* Additional reinforcement provided in the form of rings or saddles shall be of reasonably constant width.

Example

A 10 in. nominal diameter pipe has design conditions of 650°F and 400 psig. It is made from seamless material to specification ASTM A53 Grade B sch 20. The corrosion allowance is 0.03 in. It has a 4 in. nominal diameter branch, sch 40 of the same material. What are suitable dimensions for the reinforcement if it is to be made from a plate of equal quality to that of the pipe material?

We start off by calculating the minimum thicknesses required for both the 10 in. header and the 4 in. branch from the basic equation:

$$t_{pressure} = \frac{PD_o}{2(SE_q + PY)} \tag{2.14}$$

Allowable stress for ASTM A53 Grade B at 650°F $= 15{,}000 \text{ lb/in.}^2$
From Table 2.1, factor Y $= 0.4 \text{ (below 900°F)}$

For header, $t_{pressure} = \dfrac{400 \times 10.75}{2(15{,}000 \times 1.0 + 400 \times 0.4)}$ $= 0.1418 \text{ in.}$

For branch, $t_{pressure} = \dfrac{400 \times 4.5}{2(15{,}000 \times 1.0 + 400 \times 0.4)}$ $= 0.0593 \text{ in.}$

Then:

Minimum thickness of 10 in. sch 20 = 0.219 in.;

$$\text{Excess} = 0.219 - 0.1418 - 0.03$$
$$= 0.0472 \text{ in.}$$

*Multiple opening reinforcement B31.3, Section 304.3.3(b).

Minimum thickness of 4 in. sch 40 = 0.207 in.;

$$\text{Excess} = 0.207 - 0.0593 - 0.03$$
$$= 0.1177 \text{ in.}$$

The minimum thicknesses above are the nominal schedule dimensions less $12\frac{1}{2}\%$ mill tolerance (MT) allowed by the standards.

$$\text{Effective length, } d_1 = 4.5 - 2(0.1177) = 4.2646 \text{ in.}$$
$$d_2 = d_1 = 4.2646 \text{ in.}$$

The L_4 is the minimum of $2.5(\bar{T}_h - c)$ or $2.5(\bar{T}_b - c) + T_r$, that is, it is the minimum of 2.5×0.22 or $2.5 \times 0.207 + 0.25$. (Assume $\frac{1}{4}$ in. reinforcement.)
Clearly, the first condition governs so that $L_4 = 0.55$ in.
Required area $= t_{min} \times d_1 = 0.1418 \times 4.2646 \quad = 0.6047$ in.^{2}
Compensation area available from header,
$$A_2 = (2d_2 - d_1)(\text{excess thickness})$$
$$= 4.2646 \times 0.0472 = 0.2012 \text{ in.}^2$$
Compensation area available from branch,
$$A_3 = (2L_4)(\text{excess thickness})$$
$$= 1.1 \times 0.1177 = 0.1294 \text{ in.}^2$$
Total compensation available without reinforcing pad $= 0.3306$ in.^{2}
Cross-section area of pad required $= (0.6047 - 0.3306)/2 \quad = 0.1370$ in.^{2}

This results in a ring with $11\frac{3}{4}$ in. outside diameter, $\frac{9}{16}$ in. wide, and $\frac{1}{4}$ in. thick. Our neglect of the area of the weld fillets makes no difference in practice. It must be pointed out, however, that for a service of this severity a weldolet would be preferred. For more example problems, see Appendix Table A5.

EXERCISES

1. Calculate internal pressure design thickness for 8 in. carbon steel A106 Grade B pipe under 420 psig at 800°F. If mill tolerance (MT) = 12.5% and corrosion allowance is 0.05 in. select commercially available thickness.

2. Calculate maximum allowable pressure which can be held in a 12 in. standard weight A53 Grade B pipe at 725°F. Assume usual MT and 0.1 in. for corrosion allowance.

3. Select the commercially available thickness to hold 500 psig at 700°F. Pipe is 12 in. A106 Grade A material, MT is 12.5%, and corrosion allowance is 0.06 in.

REFERENCE

1. Roarke, R. J. "Formulas for Stress and Strain."

CHAPTER THREE

PIPE SPAN CALCULATION

The maximum allowable spans for horizontal piping systems are limited by three main factors: bending stress, vertical deflection, and natural frequency. By relating natural frequency and deflection limitation, the allowable span can be determined as the lower of the calculated support spacings based on bending stress and deflection.

SPAN LIMITATIONS

The formulation and equation obtained depend upon the end conditions assumed. By assuming a straight pipe beam, simply supported at both ends, Eqs. 3.1 and 3.2 are obtained (reference 1). This end condition gives higher stress and sag and therefore results in a conservative span.

$$L = \sqrt{\frac{0.33\,ZS_h}{w}} \qquad \text{based on limitation of stress} \qquad (3.1)$$

$$L = \sqrt[4]{\frac{\Delta EI}{22.5\,w}} \qquad \text{based on limitation of deflection} \qquad (3.2)$$

The end conditions can be also assumed as a mean between a uniformly loaded beam simply supported at both ends and a uniformly loaded beam with both ends fixed. With this assumption (reference 2) Eqs. 3.3 and 3.4 are obtained:

$$L = \sqrt{\frac{0.4\,ZS_h}{w}} \qquad \text{based on limitation of stress} \qquad (3.3)$$

$$L = \sqrt[4]{\frac{\Delta EI}{13.5\,w}} \qquad \text{based on limitation of deflection} \qquad (3.4)$$

where L = allowable pipe span, feet
$\quad$ Z = modulus of section of pipe, in.^{3}
$\quad$ S_h = allowable tensile stress for the pipe material at design temperature psi (known as allowable hot stress)
$\quad$ w = total weight of pipe, lb/ft
$\quad\quad$ = metal weight + content weight + insulation weight
$\quad$ Δ = allowable deflection or sag, inches
$\quad$ I = area moment of inertia of pipe, in.^{4}
$\quad$ E = modulus of elasticity of the pipe material at design temperature, psi (known as hot modulus of elasticity)

The exceptions are:

1. The piping is in a static state, except for movement induced by temperature changes. Effects of pulsation, vibration, sway, or earthquake are not taken into account.
2. Concentrated loads similar to valves are not considered in Eqs. 3.1 through 3.4.

NATURAL FREQUENCY

For most refinery piping a natural frequency of about 4 cps is sufficient to avoid resonance in nonpulsating pipe lines. However, the natural frequency f_n in cycles per second is related to the maximum deflection Δ in inches by:

$$f_n = \frac{1}{2\pi}\sqrt{\frac{g}{\Delta}} = \frac{3.12}{\sqrt{\Delta}} \tag{3.5}$$

where g = acceleration due to gravity, 386 in./sec^2 (32.12 ft/sec^2).

$\quad$ Therefore the natural frequency for a simple beam corresponding to 1.00 in. sag is 3.12 cps. One of the reasons for limiting the deflection is to make the pipe stiff enough with high enough natural frequency to avoid large amplitude under any small disturbing force. Although this may seem too low, in practice the natural frequency will be higher because (1) end moments, neglected here, will raise the frequency by more than 15%; (2) the critical span is usually limited by stress and is rarely reached; and (3) the piping weight assumed is often larger than the actual load.

$\quad$ By relating natural frequency and deflection limitations, the maximum span is thus determined by the smaller valves obtained by Eqs. 3.3 and 3.4.

$\quad$ The calculated span is then multiplied by the span reduction factor. Figure 3.1 shows different piping arrangements and span reduction factor f' (reference 3). As can be seen, span reduction factor is less than 1.0.

$\quad$ Assuming that the piping is simply supported at both ends and the valve is

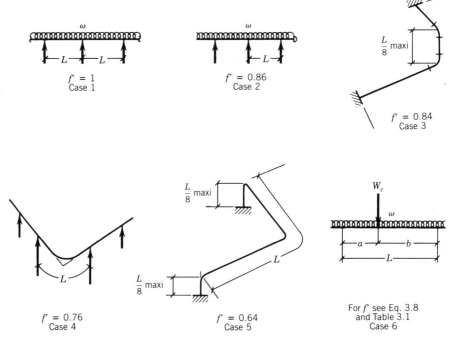

FIGURE 3.1 Piping span reduction factor case 1 through case 6.

located at midspan (in case 6 of Fig. 3.1, $a = b = L/2$), Eqs. 3.6 and 3.7 can be derived:

$$\text{Bending stress} = \frac{1.5\,wL^2 + 3\,W_c L}{Z} \tag{3.6}$$

$$\text{Deflection} = \frac{22.5\,wL^4 + 36\,W_c L^3}{EI} \tag{3.7}$$

where W_c = concentrated weight similar to the valve in pounds.

As can be easily seen, Eqs. 3.6 and 3.7 may be used to calculate actual bending stress and deflection when the span is eigher known or assumed.

To calculate allowable spans for piping with concentrated weight similar to valves anywhere along its span (case 6, Fig. 3.1), span reduction factors may be used. For a beam with fixed ends the span reduction factor is obtained (reference 4) by comparing the moment acting at the support with the moment obtained from only uniformly distributed weight and is given by:

$$f' = \left[\frac{1}{1 + 12\alpha\beta(1 - \beta)^2} \right]^{1/2} \tag{3.8}$$

TABLE 3.1   Span Reduction Factor f' for Valve Location (using Eq. 3.8)

α	$\beta = \dfrac{a}{L}$							
	0.05	0.1	0.15	0.2	0.25	0.3	0.4	0.5
0.1	0.97	0.95	0.94	0.93	0.92	0.92	0.92	0.93
0.2	0.95	0.92	0.89	0.87	0.86	0.86	0.86	0.88
0.5	0.93	0.82	0.78	0.75	0.74	0.73	0.73	0.76
0.75	0.845	0.76	0.71	0.68	0.655	0.655	0.66	0.68
1.0	0.81	0.71	0.66	0.63	0.61	0.60	0.61	0.63
1.25	0.775	0.67	0.615	0.585	0.565	0.56	0.565	0.54
1.50	0.74	0.64	0.58	0.55	0.53	0.52	0.53	0.55
1.75	0.715	0.605	0.555	0.525	0.505	0.495	0.495	0.525
2.00	0.69	0.58	0.53	0.50	0.48	0.47	0.47	0.5
2.50	0.65	0.54	0.49	0.45	0.44	0.43	0.43	0.46
4.00	0.56	0.45	0.40	0.37	0.36	0.35	0.36	0.38
5.00	0.52	0.41	0.37	0.34	0.33	0.32	0.32	0.34

The left margin label for α column: $\alpha = \dfrac{W_c}{w(a+b)}$

where

$$\alpha = \frac{W_c}{w(a+b)}, \qquad \beta = \frac{a}{L}$$

Table 3.1 gives valves of f' for different valves of α and β.

DRAINAGE

It is often necessary to install pipe systems so they will drain by gravity, preferably in the direction of normal flow. To achieve drainage each span must be pitched so that the outlet will be lower than the maximum sag of the pipe. The pitch of pipe spans is the ratio between the drop in elevation and the length of span. It is called "average gradient" and is expressed in inches per foot.

Gradient Check for Drainage

$$\text{Average gradient } G = \frac{\text{drop in elevation}}{\text{span}} \text{ in./ft} \qquad (3.9)$$

TABLE 3.2   Common Pipe Insulation Materials
(Mass Type)

Pipe Insulation Type	Density (lb/ft³)
Calcium silicate	12.25
Foam glass	8.50
Polyurethane	2.00
Fiber glass	3.25
Polystyrene	2.00

The condition for good drainage is:

$$G \leq \frac{4(\text{maximum deflection})}{\text{span}} \tag{3.10}$$

In calculating modulus of section and area moment inertia of piping, corrosion allowance may be included, which will result in a slightly higher span.

Table 3.2 gives the common piping mass-type insulation materials. The other type is known as reflective and is used inside reactor buildings of nuclear plants (reference 7).

To illustrate the use of the preceding equations the following Example Problems are worked out.

Example Problem 1

Calculate the allowable span for a 10 in. pipe with standard wall operating at 400°F. The material of the piping is carbon steel A106 Grade B. The pipe is filled with crude oil whose specific gravity is 1.2; it has a 2 in. thick calcium silicate insulation of density 11 lb/ft³. Metal weight, content weight, and insulation weight may also be obtained from any standard table. Assume that the maximum deflection allowed is $\frac{5}{8}$ in.

$$\text{Self weight of pipe} = \frac{\pi}{4}(\text{OD}^2 - \text{ID}^2)(\text{density of steel})(\text{length})$$

$$= \frac{\pi}{4}(10.75^2 - 10.02^2)(0.283)(12)$$

$$= 40.44 \text{ lb/ft}$$

$$\text{Content weight} = \frac{\pi}{4}(ID)^2(\text{length})(\text{density})$$

$$= \frac{\pi}{4}(10.02)^2(12)\frac{62.4}{(12)^3}$$

$$= 34.17 \text{ lb/ft}$$

$$\text{Insulation weight} = \frac{\pi}{4}(\text{OD of insulation} - \text{OD})^2(\text{length})(\text{density})$$

$$= \frac{\pi}{4}(14.75^2 - 10.75^2)(12)\left(\frac{11}{12^3}\right)$$

$$= 6.12 \text{ lb/ft}$$

Total weight of the pipe $w = 40.44 + 34.17 + 6.12$ lb/ft

$$= 80.73 \text{ lb/ft}$$

Using Eq. 3.3 based on limitation of stress:

$$\text{Span } L = \sqrt{\frac{0.4 Z S_h}{w}}$$

$Z =$ section modulus $= 29.9$ in.^{3}

$S_h =$ allowable stress of pipe material for design temperature

$\quad = 22{,}900$ psi for carbon steel A106 Grade B at 400°F per B31.3 code (Appendix Table A1)

$$\text{Span } L = \sqrt{\frac{0.4(29.9)22{,}900}{80.73}} = 58.2 \text{ ft}$$

Using Eq. 3.4 based on limitation of allowable deflection of $\frac{5}{8}$ in.:

$$\text{Span } L = \sqrt[4]{\frac{\Delta E I}{13.5 w}} = \sqrt[4]{\frac{\frac{5}{8}(27 \times 10^6)160.7}{13.5(80.73)}} = 39.7 \text{ ft}$$

$E =$ Young's modulus (psi) at 400°F for carbon steel with carbon content 0.3% or less

$\quad = 27.0 \times 10^6$ psi

$I =$ area moment of inertia of pipe $= 160.7$ in.^{4}

By selecting the smaller of 58.2 ft and 39.7 ft as span, the span is 39.7 ft.

Example Problem 2

Calculate also the maximum allowable span in the following case (using basic information from *Example* 1).
(a) If 1 in. static deflection was allowed.
(b) If the material of the pipe was stainless steel A312 TP 314.
(c) If the piping material was aluminum seamless B241 Grade 6061 T6.
(d) If the piping material was red brass, seamless B43 (commercial brass 66 Cu–343 n).
(e) If nickel piping was used (Ni Cu, specification B165, P No. 42, Grade 400, hot annealed).

Case (a): If 1 in. static deflection is allowed:

$$L = \sqrt[4]{\frac{1(27 \times 10^6)160.7}{13.5(80.73)}}$$

$$= 44.6\,\text{ft}$$

Case (b): If the material of the pipe was stainless steel A312 TP 304 (18 Cr–8 Ni pipe):

$$L = \sqrt{\frac{0.4 Z S_h}{w}}$$

$$= \sqrt{\frac{0.4(29.9)18,700}{80.73}} = 52.6\,\text{ft}$$

$$L = \sqrt[4]{\frac{\frac{5}{8}(26.6)10^6(160.7)}{13.5(80.73)}} = 39.56\,\text{ft}$$

Required span, $L = 39.56\,\text{ft}$

Case (c): If the piping material was aluminum seamless B241 Grade 6061 T6 at 400°F:

$$L = \sqrt{\frac{0.4(29.9)5600}{80.73}} = 28.8\,\text{ft}$$

$$L = \sqrt[4]{\frac{\frac{5}{8}(8.7 \times 10^6)160.7}{13.5(80.73)}} = 30.0\,\text{ft}$$

Required span, $L = 28.8\,\text{ft}$

Case (d): If the piping material was red brass seamless B43 (commercial brass 66 Cu–343):

$$L = \sqrt{\frac{4(29.9)5000}{80.73}} = 27.2 \text{ ft}$$

$$L = \sqrt[4]{\frac{\frac{5}{8}(13 \times 10^6)160.7}{13.5(80.73)}} = 33 \text{ ft}$$

Required span, $L = 27.2$ ft

Case (e): If the material was Ni–Cu, specification B165, P No. 42 Grade 400, hot annealed:

$$L = \sqrt{\frac{4(29.9)13,200}{80.73}} = 44.2 \text{ ft}$$

$$L = \sqrt[4]{\frac{\frac{5}{8}(20.6)10^6(160.7)}{13.5(80.73)}} = 37.1 \text{ ft}$$

Required span, $L = 37.1$ ft

Table of span: To provide the reader with a quick reference values of span, Table 3.3a and Table 3.3b are presented.

The following assumptions were made:

1. Pipe material is carbon steel A53 Grade A. Table 3.3a applies conservatively to all other steels.
2. Temperature ranges from zero to 650°F. At 650°F, $S_h = 12,000$ psi. Modulus of elasticity $E_h = 25.2 \times 10^6$ psi from the piping code.
3. Specific gravity of fluid is 1.0 (water).
4. Density of insulation is 11 lb/ft³.
   Thickness of insulation is $1\frac{1}{2}$ in. for pipe sizes 1–4 in.
   $\qquad\qquad\qquad\qquad\qquad$ 2 in. for pipe sizes 6–14 in.
   $\qquad\qquad\qquad\qquad\qquad$ $2\frac{1}{2}$ in. for pipe sizes 16–24 in.
5. The pipe was treated as a horizontal beam, supported at both ends, carrying a uniform load equal to the combined weight of metal weight, water, and insulation.
6. The maximum static deflection was 1 in. and natural frequency was 3.12 cps.
7. The maximum bending stress was equated to allowable weight stress equal to half the allowable hot stress S_h.

TABLE 3.3a Maximum Spans of Horizontal Pipe Lines (ft) (select smaller of L and L')[a]

Pipe Size (in.)

		1	1½	2	3	4	6	8	10	12	14	16	18	20	24
Schedule 10	L	13	15	17	20	22	25	29	30	32	37	38	39	39	41
	L'	13	16	18	21	24	28	31	34	37	41	42	44	46	48
Schedule 20	L							33	35	36	39	41	42	45	47
	L'							33	37	39	42	44	46	49	52
Schedule 30	L							34	37	39	42	43	46	49	52
	L'							34	38	41	43	45	48	51	55
Standard	L	13	16	18	23	26	31	35	38	41	42	43	44	45	47
	L'	13	16	18	23	26	31	35	38	41	43	45	47	49	52
Schedule 40	L	13	16	18	23	26	31	35	38	41	43	46	49	51	56
	L'	13	16	18	23	26	31	35	38	42	44	45	50	52	57
Schedule 60	L							36	40	43	46	49	52	55	60
	L'							35	39	43	45	48	51	54	59

		1	1½	2	3	4	6	8	10	12	14	16	18	20	24
								Pipe Size (in.)							
Extra Strong	L	13	17	19	24	27	33	37	41	43	44	46	48	49	51
	L'	13	17	19	23	26	32	36	40	43	44	46	49	51	54
Schedule 80	L	13	17	19	24	27	33	37	42	46	48	52	55	58	63
	L'	13	17	19	23	26	32	36	40	44	46	50	52	55	61
Schedule 100	L							38	43	47	49	53	56	59	65
	L'							37	41	45	47	50	53	56	61
Schedule 120	L					28	34	39	44	48	51	54	57	61	67
	L'					27	32	37	41	45	47	51	54	57	62
Schedule 140	L					28	34	40	44	49	51	54	58	61	67
	L'					27	33	37	42	45	48	51	54	57	62
Schedule 160	L	13	17	20	25	29	35	40	45	49	51	55	58	62	68
	L'	13	17	19	23	27	33	37	42	45	48	51	54	57	63

[a]Span L was calculated using Eq. 3.1, with limiting bending stress of S_h divided by 2.
Span L' was determined using Eq. 3.2 with limiting static deflection of 1 in.

TABLE 3.3b   Calculation Factors (C_1, C_2, and C_3) for Spans[a]

If the allowable stress S_h is	2,000	4,000	6,000	8,000	10,000	12,000	14,000	16,000	18,000	20,000
Multiply the span L by C_1 =	0.408	0.577	0.707	0.816	0.913	1.000	1.080	1.155	1.225	1.291
If the allowable deflection (in.) Δ is	$\frac{1}{8}$	$\frac{1}{4}$	$\frac{3}{8}$	$\frac{1}{2}$	$\frac{5}{8}$	$\frac{3}{4}$	$\frac{7}{8}$	1	$1\frac{1}{4}$	$1\frac{1}{2}$
Multiply the span L' by C_2 =	0.595	0.707	0.782	0.841	0.883	0.930	0.967	1.000	1.057	1.106
If the minimum allowable frequency f is	3.12	4	5	6	7	8	9	10	15	20
Multiply the span L' by C_3 =	1.000	0.883	0.790	0.720	0.668	0.625	0.589	0.559	0.456	0.395

[a] Span L was calculated using Eq. 3.1, with limiting bending stress of S_h divided by 2.
Span L' was determined using Eq. 3.2 with limiting static deflection of 1 in.

For other values of allowable stress, deflection, and natural frequency, the span values given in Table 3.3a need to be multiplied by span calculation factors (given in Table 3.3b) C_1, C_2, and C_3.

Values in Table 3.3b were arrived at as follows:

1. For any other allowable stress S_h, the maximum span is $C_1 L$, where $C_1 = (S_h/12{,}000)^{1/2}$.
2. For deflections other than 1 in., the maximum pipe span is $C_2 L'$ where $C_2 = (\Delta/L')^{1/4}$.
3. For natural frequency f other than 3.12 cps, the maximum span is $C_3 L'$, where $C_3 = (3.12/f)^{1/2}$.

These calculation factors are given in Table 3.3b for some values of S_h and f. This calculation factor should not be confused with span reduction factors given earlier in Figure 3.1.

Example

1. Using Table 3.3a, calculate the maximum span allowed for a 14 in. sch 40 pipe. (Assume $S_h = 12{,}000$ psi, $\Delta = 1$ in., and $f = 3.12$ cps.)
   Span L considering the stress from Table 3.3a = 43 ft.
   Span L' considering the deflection   = 44 ft.
   Select the smaller of the two spans, namely 43 ft.

2. Calculate the span if S_h was 10,000 psi.
   From Table 3.3b, the calculation factor is $C_1 = 0.913$, span = 0.913(43) = 39.2 ft.

3. Calculate the span if $\Delta = \frac{1}{2}$ in.
   From Table 3.3b the calculation factor is $C_2 = 0.841$, span = 0.841(44) = 37 ft.

4. Calculate the span if the pipe is connected to a compressor with speed of 8 cps.
   From Table 3.3b, calculation factor $C_3 = 0.625$, span = 0.625(44) = 27.5.

Calculation of the allowable span under dynamic loading is complicated. The conservative formula for calculating the restraint spacing (reference 5) based on stress criterion is given by:

$$L \leq 2.19 \sqrt{\frac{S_h Z}{12 \, Kw}} \tag{3.11}$$

where K = seismic coefficient depending on the peak of floor response spectra (multiple of acceleration, G).

Dynamic deflection criterion (reference 4) can be used to calculate the allowable span under dynamic loading.

For a simply supported single span beam, the maximum deflection by taking one mode is given as:

$$\text{Maximum } \Delta = \frac{4m}{\pi^5} \frac{L^4}{EI} A_{an} \tag{3.12}$$

where m = pipe mass/foot
E = modulus of elasticity, psi
I = moment of inertia, in.^{4}
A_{an} = seismic acceleration of pipe, ft/sec^2.

GUIDE SPACING FOR WIND LOADING

Table 3.4 gives maximum spacing of guides for vertical piping.

Table 3.5 gives suggested pipe support spacing (span) as per ASME Nuclear Code, Section III, Division 1, Subsection NF-3133.1-1.

TABLE 3.4   Maximum Spacing of Guides

Nominal Pipe Size (in.)	Guide Spacing (ft)
1	22
$1\frac{1}{2}$	23
2	24
3	27
4	29
6	33
8	37
10	41
12	45
14	47
16	50
18	53
20	56
24	60

Notes:

1. Guides should be kept about 40 pipe diameters clear of corners or loops.

2. Use of pipe guides on hot lines must be investigated to assure that no higher forces or stresses are transmitted to the piping system due to the location of the guide.

3. Calculation of wind loads on pipes is given in reference 6.

TABLE 3.5 **Suggested Pipe Support Spacing**

Nominal Pipe Size (in.)	Suggested Maximum Span (ft)	
	Water Service	Steam, Gas, or Air Service
1	7	9
2	10	13
3	12	15
4	14	17
6	17	21
8	19	24
12	23	30
16	27	35
20	30	39
24	32	42

Notes:
1. Suggested maximum spacing between pipe supports for horizontal straight runs of standard and heavier pipe at maximum operating temperatures of 750°F.
2. Does not apply where span calculations are made or where there are concentrated loads between supports such as flanges, valves, and specialities.
3. The spacing is based on a maximum combined bending and shear stress of 1500 psi and insulated pipe filled with water or the equivalent weight of steel pipe for steam, gas, or air service and the pitch of the line is such that a sag of 0.1 in. between supports is permissible.

DESIGN RULES FOR PIPE SUPPORTS

Spacing of Piping Supports

Supports for piping with longitudinal axis in approximately a horizontal position shall be spaced to prevent excessive shear stresses resulting from sag and bending in the piping with special consideration given when components such as pumps and valves that impose concentrated loads. The suggested maximum spans for spacing of weight supports for standard weight and heavier pipe are given in Table 3.5.

EXERCISES

1. Calculate the maximum allowable span for a 16 in. standard weight pipe filled with oil whose specific gravity is 0.95. The pipe is insulated with 3 in.

thick calcium silicate with density 12.25 lb/ft^3. The material of the piping is carbon steel A106 Grade B and the temperature of the oil is 600°F. Assume the maximum deflection allowed is 1 in.

2. Calculate the span if a valve weighing 1050 lb was located at the midspan of Exercise 1.

3. Calculate the span if a valve weighing 1050 lb was located at one-third span distance from one support of Exercise 1.

4. Calculate the span if the pipe considered in Exercise 1 has a 90 degree elbow between the supports.

5. Calculate the static deflection in a 10 in. sch 80 stainless steel pipe filled with water and with 3 in. of fiber glass insulation.

REFERENCES

1. Barc, W. et al., *Pipe Supports for Industrial Piping Systems*, Procon Inc., 1963.
2. Fluor Design Guides and Q. Truong. Seminar on Piping Systems, A&M University, Texas.
3. DMI, Inc. *Design Standards*.
4. Niyogi, B. K. "Simplied Seismic Analysis Methods for Small Pipes." ASME 78-PVP-43.
5. Stevenson et al., "Seismic Design of Small Diameter Pipe and Tubing for Nuclear Power Plants," Paper #314, Fifth World Conference of Earthquake Engineering, Rome, 1973.
6. ANSI Standard A58.1 "Wind Loads for Buildings and Other Structures."
7. Wilkes, Gordon B. "Heat Insulation," Wiley, NY.

CHAPTER FOUR

ANSI PIPING CODES
AND ASME CODES

The ANSI Piping Codes and ASME Pressure Vessel codes give guidelines for piping design. In general, the latest revision of the code should be used. In the design of Nuclear Power Plants Piping, the code of record, which is not necessarily the latest revision, for a specific plant can be used.

Codes related to piping design include:

1. ANSI B31.1, Power Piping (reference 1)
2. ANSI B31.3, Chemical Plant and Petroleum Refinery Piping (reference 2)
3. ANSI B31.4, Liquid Transportation Piping (reference 3)
4. ANSI B31.8, Gas Transportation Piping (reference 4)
5. ASME Section III, Nuclear Components Design (reference 5)*

Subsection NA	General Appendix, Material Properties
Subsection NB	Class 1 piping (high energy piping)
Subsection NC	Class 2 piping
Subsection ND	Class 3 piping
Subsection NF	Support design

Nuclear components design is treated in Chapter 10.

* See Chapter 6 (flanges) and Chapter 10 (Dynamics Analysis) for explanation of service levels.

INTERNAL PRESSURE AND
LONGITUDINAL STRESSES

Code allowable stresses are designed to prevent failure of the piping systems. Two types of failure that the piping should be protected against are:

1. Direct overstress or failure due to pressure, weight, wind, earthquake, and other primary loads.
2. Fatigue or distortion due to displacement strains (generally thermal effects) which are secondary loads.

The limits of calculated stresses caused by sustained loads and displacement strains are:

1. *Internal Pressure Stresses*:   Stresses due to internal pressure is considered safe when the pipe wall thickness and any reinforcement are adequate. (See thickness calculation in Chapter 2.)
2. *Longitudinal Stresses* (S_L):   The sum of longitudinal stresses resulting from pressure, weight, and other sustained loadings shall not exceed the basic allowable stress for material at maximum metal temperature S_h. Pipe thickness used in the calculation of S_L must be reduced by allowances such as corrosion, erosion, manufacturing tolerance, and grove depth.
3. *Allowable Stress Range for Displacement Stresses*:   The allowable stress range S_A is given by:

$$S_A = f(1.25 S_C + 0.25 S_h) \qquad (4.1)$$

where S_C = basic allowable stress for the material at minimum (cold) metal temperature, psi
   S_h = hot stress, psi
   f = stress reduction factor for cyclic conditions for the total number of full temperature cycles over expected life.
Table 4.1 gives values of stress range reduction factors, f.

TABLE 4.1   Stress Range Reduction Factors

Number of Cycles	Factor f
7,000 and less	1.0
7,000 to 14,000	0.9
14,000 to 22,000	0.8
22,000 to 45,000	0.7
45,000 to 100,000	0.6
Over 100,000	0.5

TABLE 4.2 Comparison of Allowable Stresses and Yield Stress for Seamless Pipe (ksi)[a]

		Code	Metal Temperature °F															σ_y
			−20 to −100	200	300	400	500	600	650	700	750	800	850	900	950	1000	
Carbon Steel	A106 Grade B	B31.3	20	20	20	20	18.9	17.3	17.0	16.8	13.0	10.8	–	–	–	–	30
		B31.1	14.4	14.4	14.4	14.4	14.4	14.4	14.4	14.4	13	10.8	–	–	–	–	
		Sec. III Class 2	15	15	15	15	15	15	15	14.3	–	–	–	–	–	–	35
Low and Intermediate Alloy Steel	5% Cr-½ Mo A335 Grade P5	B31.3	20	18.1	17.4	17.2	17.1	16.8	16.6	16.3	13.2	12.8	12.1	10.9	8.0	5.8	30
		B31.1	15	15	15	15	14.5	14	13.7	13.4	13.1	12.8	12	10.4	7.6	5.6	
		Sec. III Class 2	15	15	14.5	14.4	14.4	14.2	13.9	13.7	–	–	–	–	–	–	30
Stainless Steel	18 Cr-8 Ni A312 TP 304 P No. 8	B31.3	20	20	20	18.7	17.5	16.4	16.2	16	15.6	15.2	14.9	14.6	14.4	13.8	30
		B31.1	18.8	15.7	14.1	13	12.2	11.4	11.2	11.1	10.8	10.6	10.4	10.2	10	9.8	
		Sec. III Class 2	17.5	16.6	16.1	15.5	15	14.5	14.3	14.1	–	–	–	–	–	–	40

[a] For other material values, see Appendix Table A3.

When S_h is greater than the calculated value of S_L, the difference between them may be added to the term $0.25 S_h$ in Eq. 4.1. In this case the allowable stress range becomes:

$$S_A = f[1.25(S_c + S_h) - S_L] \qquad (4.2)$$

Appendix Table A3 gives values of cold stress S_c and hot stress S_h for piping materials from B31.3 piping code.

Representative values of S_c and S_h for carbon steel, alloy steel A335 5 Cr-$\frac{1}{2}$ Mo, and for stainless steel A312 TP 304 are given in Table 4.2 from B31.1 (reference 1), B31.3 (reference 2), and Section III (Class 2 materials subsection NC, reference 5). As can be seen, B31.3 code gives higher allowable stress, whereas Section III, Class 2 materials are allowed higher yield stress. Appendix Table A3 gives B31.3 values for the most common materials. For other codes, appropriate references should be used in the actual design.

Calculation of allowable stress range S_A using Eq. 4.1 is frequently encountered. Three examples are given here to show the calculation of S_A.

Example

1. A pipe is fabricated of seamless carbon steel to specification ASTM A106 Grade B. The design temperature is 700°F. What is the allowable expansion stress range? Refer to ANSI B31.3 (latest edition) to find the value of S_c and S_h stress at minimum temperature to 100°F (i.e., S_c) = 20,000 lb/in.^{2} Stress at 700°F (i.e., S_h) = 16,800 lb/in.^{2} (Appendix Table A3). In the absence of any reason for taking a lower value assume $f = 1.0$; then $S_A = 1.0(1.25 \times 20,000 + 0.25 \times 16,800) = 29,000$ lb/in.^{2}

2. A pipe supplies steam to a jacketed process vessel that is operated on a batch process with a 4 hour cycle. The steam temperature is 200°F and the material of the pipe is a seamless low and intermediate alloy steel pipe, ASTM A335 5 Cr-$\frac{1}{2}$ Mo. If the installation is operated continuously and the design life is to be 12 years, what is the allowable stress range for thermal stresses in the pipe?

    Allowable stress (cold) = S_c = 20,000 lb/in.^{2} (Appendix Table A3)

    Allowable stress (325°F) = S_h = 18,100 lb/in.^{2}

    Number of cycles = $\frac{24}{4} \times 365 \times 12 = 26,280$

    $f = 0.7$ (for 22,000–45,000 cycles)   (see Table 4.1)

    $S_A = f(1.25 S_c + 0.25 S_h)$

    $S_A = 0.7(1.25 \times 20,000 + 0.25 \times 18,100) = 20,667$ lb/in.^{2}

3. A line in a relief system attains a temperature of −90°F when the relief

valve lifts. The material is stainless steel pipe A312 TP 304 (18 Cr–18 Ni pipe). What is the allowable expansion stress range?

Two things to note are:

1. Because relief valves do not operate very frequently, we will be justified in assuming that the pipe will experience less than 7000 cycles of stress. Therefore $f = 1.0$.
2. The fact that the range from ambient to operating temperatures is negative makes no difference. It is the temperature change that matters.

For ASTM A312 TP 304 seamless pipe, the allowable stress is (minimum to 100°F)

$$S_c = S_h = 20,000$$

$$S_A = f(1.25 S_c + 0.25 S_h) \text{ psi}$$

$$S_A = 1.0(1.25 \times 20,000 + 0.25 \times 20,000) = 30,000 \text{ psi.}$$

PETROLEUM REFINERY PIPING CODE REQUIREMENTS FOR FORMAL ANALYSIS*

No formal analysis of adequate flexibility is required in systems which:

1. Are duplicates of successfully operating installations or replacements without significant change of systems with a satisfactory service record
2. Can be readily adjudged adequate by comparison with previously analyzed systems.
3. Are of uniform size, have no more than two points of fixation and no intermediate restraints, and satisfy Eq. 4.3:

$$\frac{Dy}{(L - U)^2} \leq \frac{30 S_A}{E_a} \qquad (4.3)$$

where D = nominal pipe size, inches
$\quad\quad y$ = resultant of total displacement strains to be absorbed by the piping system, inches
$\quad\quad U$ = Anchor distance, straight line distance between anchors, feet
$\quad\quad L$ = developed length of piping between anchors, feet
$\quad\quad S_A$ = allowable stress range, psi (include stress range reduction factor f

---

* From ASME/ANSI B31.3, subsection 319.4.1, Requirements for Analysis.

where more than 7000 cycles of movement are anticipated during life of installation).

E_a = modulus of elasticity of the piping material in the cold condition, psi

Because no general proof can be offered that Eq. 4.3 will always be conservative, caution should be exercised in applying it to abnormal configurations (unequal leg U-bends), to large diameter thin-wall pipe (stress intensification factors of the order of five or more), or to conditions where extraneous motions, other than in the direction connecting the anchor points, constitute a large proportion of the expansion duty.

User must be aware that compliance with Eq. 4.3 does not ensure that the terminal reactions will be satisfactory. A value of 0.03 may be assumed for the right-hand side in Eq. 4.3 if enough information was not available. (Eq. 4.3 does not include weight effect.)

Example

4. Check if formal analysis is necessary in the piping arrangement given in Figure 4.1 using Eq. 4.3.

    The diameter is 10 in., temperature is 300°F, coefficient is 0.023 in./ft of pipe for carbon steel A106 Grade B (see Appendix Table A1).

    The expansion in each direction and terminal movement is:

$\Delta x = 40(0.023) = 0.92$ in.

$\Delta y = (50 - 10)(0.023) + (2 - 1) = 1.92$ in.

$\Delta z = 15(0.023) = 0.345$ in.

$y = \sqrt{0.92^2 + 1.92^2 + 0.119^2} = 2.13.$

D = nominal pipe size = 10 in.

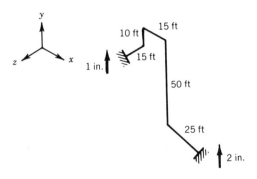

FIGURE 4.1 Formal analysis requirement.

E_a = cold modulus for carbon steel = 27.9×10^6 psi (Appendix Table A2)

L = developed length = $15 + 10 + 15 + 50 + 25 = 115$ ft

U = anchor distance = 58.5 ft (straight line distance between anchors)

$S_A = 1(1.25 S_c + 0.25 S_h) = 1.25(20,000) + 0.25(20,000)$
$$= 30,000 \text{ psi}$$

Equation 4.3 states that formal analysis is not necessary if:

$$\frac{Dy}{(L-U)^2} \leq \frac{30 S_A}{E_a} \tag{4.3}$$

$$\frac{Dy}{(L-U)^2} = \frac{10(2.13)}{(115-58.5)^2} = \frac{10(2.13)}{3192.25} = 0.00668$$

$$\frac{30 S_A}{E_a} = \frac{30(30,000)}{27.9 \times 10^6} = \frac{0.1}{27.9} = 0.0322$$

Because $Dy/(L-U)^2 \leq 30 S_A/E_a$, no formal analysis is necessary from the thermal flexibility point of view.

INPLANE AND OUTPLANE BENDING MOMENTS

The B31.3 code defines inplane and outplane bending moments, which are shown in Figures 4.2 and 4.3. After application of the inplane bending moment

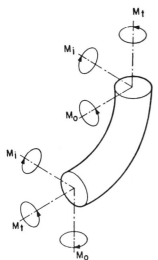

FIGURE 4.2 Inplane and outplane moments in a bend (ANSI/ASME B31.3).

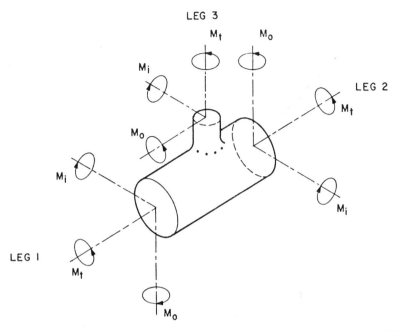

FIGURE 4.3   Inplane and outplane moments in branch connection (ANSI/ASME B31.3).

M_i, the bend or branch connection still remains in the original plane. But when outplane bending moment M_o was applied, the bend or branch connection goes out of the original plane. The torsional moment about the axis of the pipe is denoted by M_t. Power Piping Code (B31.1) and Nuclear Code (ASME Sec. III) do not differentiate between inplane and outplane bending moments. See Table 10.1 for Nuclear Code equations.

STRESS INTENSIFICATION FACTORS

Piping auxiliaries like bends (e.g., elbows, miter bends) and branch connections (e.g., welding tee, fabricated tees) have flexibility characteristic h, flexibility factor k, and stress intensification factors (SIF). In this area most codes including British standard BS3351 use the work done by Markl (reference 6). Table 4.3a (reproduced from Appendix D, ANSI B31.3, 1980 revision) gives equations for calculating values for h, k, inplane SIF i_i, and outplane SIF i_o. Note that other codes do *not* allow the use of lower value for outplane SIF ($0.75/h^{2/3}$) compared with higher value of $0.9/h^{2/3}$ for inplane SIF.

TABLE 4.3a Flexibility Factor k and Stress Intensification Factor i

Description	Flexibility Factor k	Stress Int. Factor[1,8] Outplane i_o	Stress Int. Factor[1,8] Inplane i_i	Flexibility Characteristic h	Sketch
Welding elbow[1,2,3,6,9 a] or pipe bend	$\dfrac{1.65}{h}$	$\dfrac{0.75}{h^{2/3}}$	$\dfrac{0.9}{h^{2/3}}$	$\dfrac{\bar{T}R_1}{(r_2)^2}$	$\bar{T}$, r_2, R_1 = Bend Radius
Closely spaced miter bend[1,2,3] $s < r_2(1+\tan\theta)$	$\dfrac{1.52}{h^{5/6}}$	$\dfrac{0.9}{h^{2/3}}$	$\dfrac{0.9}{h^{2/3}}$	$\dfrac{\cot\theta}{2}\dfrac{\bar{T}_s}{(r_2)^2}$	r_2, $\bar{T}$, $R_1 = \dfrac{s\cot\theta}{2}$, s, θ
Single miter bend[1,2] or widely spaced miter bend $s \geq r_2(1+\tan\theta)$	$\dfrac{1.52}{h^{5/6}}$	$\dfrac{0.9}{h^{2/3}}$	$\dfrac{0.9}{h^{2/3}}$	$\dfrac{1+\cot\theta}{2}\dfrac{\bar{T}}{r_2}$	r_2, $\bar{T}$, $R_1 = \dfrac{r_2(1+\cot\theta)}{2}$, s, θ
Welding tee[1,2,6] per ANSI B16.9 with $r_x \geq \tfrac{1}{8}D_b$ $\bar{T}_c \geq 1.5\bar{T}$	1	$\dfrac{0.9}{h^{2/3}}$	$\tfrac{3}{4}i_o + \tfrac{1}{4}$	$4.4\dfrac{\bar{T}}{r_2}$	$\bar{T}$, r_2, $\bar{T}_c$

57

TABLE 4.3a Flexibility Factor *k* and Stress Intensification Factor *i* (*Continued*)

Description	Flexibility Factor *k*	Stress Int. Factor[1,8]		Flexibility Characteristic *h*	Sketch
		Outplane i_o	Inplane i_i		
Reinforced fabricated[1,2,5] tee with pad or saddle	1	$\dfrac{0.9}{h^{2/3}}$	$\tfrac{3}{4}i_o + \tfrac{1}{4}$	$\dfrac{(\bar{T} + \tfrac{1}{2}\bar{T}_r)^{5/2}}{\bar{T}^{3/2}r_2}$	
Unreinforced[1,2] fabricated tee	1	$\dfrac{0.9}{h^{2/3}}$	$\tfrac{3}{4}i_o + \tfrac{1}{4}$	$\dfrac{\bar{T}}{r_2}$	

See Notes on page 61.

	k	$\dfrac{0.9}{h^{2/3}}$		h	
Extruded[1,2] welding tee $r_x \geq 0.05 D_b$ $T_c < 1.5\bar{T}$	1	$\dfrac{0.9}{h^{2/3}}$	$\dfrac{3}{4}i_o + \dfrac{1}{4}$	$\left(1 + \dfrac{r_x}{r_2}\right)\dfrac{\bar{T}}{r_2}$	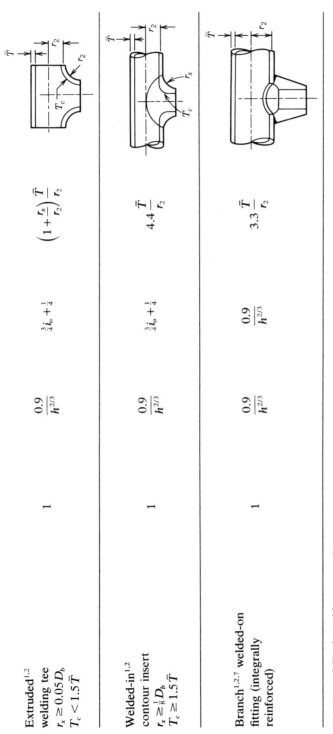
Welded-in[1,2] contour insert $r_x \geq \frac{1}{8}D_b$ $T_c \geq 1.5\bar{T}$	1	$\dfrac{0.9}{h^{2/3}}$	$\dfrac{3}{4}i_o + \dfrac{1}{4}$	$4.4\dfrac{\bar{T}}{r_2}$	
Branch[1,2,7] welded-on fitting (integrally reinforced)	1	$\dfrac{0.9}{h^{2/3}}$	$\dfrac{0.9}{h^{2/3}}$	$3.3\dfrac{\bar{T}}{r_2}$	

[a]See Notes following table on page 61.

TABLE 4.3b SIF and Flexibility Factor Chart

Description	Flexibility Factor k	Stress Intensification Factor i
Butt-welded joint, reducer, or weld neck flange	1	1.0
Double-welded slip-on flange	1	1.2
Fillet-welded joint or socket weld flange	1	1.3
Lap joint flange (with ANSI B16.9 lap joint stub)	1	1.6
Screwed pipe joint, or screwed flange	1	2.3
Corrugated straight pipe, or corrugated or crashed bend[4]	5	2.5

Source: ANSI/ASME B31.3, 1980 Editon, Appendix D.

The chart contains:

- FLEXIBILITY FACTOR FOR ELBOWS $k=1.65/h$
- FLEXIBILITY FACTOR FOR MITERS $k=1.52/h^{5/6}$
- STRESS INTENSIFICATION FACTOR $i=0.9/h^{2/3}$
- STRESS INTENSIFICATION FACTOR $i=0.75/h^{2/3}$

STRESS INTENSIFICATION FACTOR i AND FLEXIBILITY FACTOR k

Chart A

CORR. FACTOR c_1

- 1 END FLANGED $c_1=h^{1/6}$
- 2 ENDS FLANGED $c_1=h^{1/3}$

Chart B

CHARACTERISTIC h

Notes: (Applies to Table 4.3a and 4.3b)

[1] The flexibility factor k in the table applies to bending in any plane. The flexibility factors k and stress intensification factors i shall not be less than unity; factors for torsion equal unity. Both factors apply over the effective arc length (shown by heavy centerlines in the sketches) for curved and miter bends, and to the intersection point for tees.

[2] The values of k and i can be read directly from Chart A by entering with the characteristic h computed from the formulas given above. Nomenclature is as follows:

$\bar{T}$ = for elbows and miter bends the nominal wall thickness of the fitting

   = for tees, the nominal wall thickness of the matching pipe

$\dfrac{T_c}{\bar{T}}$ = the croth thickness of tees

$\bar{T}_r$ = pad or saddle thickness

θ = one-half angle between adjacent miter axes

r_2 = mean radius of matching pipe

R_1 = bend radius of welding elbow or pipe bend

r_x = crotch radius

s = miter spacing at centerline

D_b = OD of branch.

[3] Where flanges are attached to one or both ends, the values of k and i in the table shall be corrected by the factors C_1, which can be read directly from Chart B, entering with the computed h.

[4] Factors shown apply to bending. Flexibility factor for torsion equals 0.9.

[5] When $\bar{T}_r$ is $>1\frac{1}{2}\bar{T}$, use $h = 4\bar{T}/r_2$.

[6] The designer is cautioned that cast butt welded fittings may have considerably heavier walls than that of the pipe with which they are used. Large errors may be introduced unless the effect of these greater thicknesses is considered.

[7] The designer must be satisfied that this fabrication has a pressure rating equivalent to straight pipe.

[8] A single intensification factor equal to $0.9/h^{2/3}$ may be used for both i_i and i_r if desired.

[9] In large diameter thin-wall elbows and bends, pressure can significantly affect the magnitudes of k and i. To correct values from the table

divide k by: $\left[1 + 6\left(\dfrac{P}{E_c}\right)\left(\dfrac{r_2}{\bar{T}}\right)^{7/3}\left(\dfrac{R_1}{r_2}\right)^{1/3}\right]$ divide i by: $\left[1 + 3.25\left(\dfrac{P}{E_c}\right)\left(\dfrac{r_2}{\bar{T}}\right)^{5/2}\left(\dfrac{R_1}{r_2}\right)^{2/3}\right]$

Example

5. Calculate SIF and flexibility factor for 12 in. standard schedule long radius elbow.

   (a) For welding elbow,

   bend radius = R_1 = 1.5(nominal diameter) = 1.5(12) = 18 in.

   $\bar{T}$ = nominal wall thickness = 0.375 in. (see Appendix Table A4, properties of pipe) and assume that elbow and pipe have same thickness.

   $$r_2 = \text{mean radius of pipe} = \frac{OD - \bar{T}}{2} = \frac{12.75 - 0.375}{2} = 6.1875 \text{ in.}$$

   $$h = \text{flexibility characteristic} = \frac{\bar{T}R_1}{(r_2)^2} \quad \text{(equation from Table 4.3a)}$$

   $$= \frac{0.375(18)}{(6.1875)^2} = 0.176$$

   $$k = \text{flexibility factor} = \frac{1.65}{h} = \frac{1.65}{0.176} = 9.358$$

   $$i_i = \text{inplane stress intensification factor} = \frac{0.9}{h^{2/3}} = \frac{0.9}{(0.176)^{2/3}} = 2.86$$

   $$i_o = \text{outplane SIF} = \frac{0.75}{(0.176)^{2/3}} = 2.4$$

   The lower value $i_i = 0.75/h^{2/3}$ is allowed for B31.3 and B31.4. If desired, a higher value $i = 0.9/h^{2/3}$ may be used for both i_i and i_o. Chart A in Table 4.3b may be used to read i_i and i_o. For B31.1, Power Piping and Nuclear Piping, Section III, Classes 2 and 3 piping, use the higher value only.

   (b) If one end is flanged, the correction factor = $C_1 = h^{1/3} = (0.176)^{1/3} = 0.5604$.

   $$\text{Flexibility factor} = C_1\left(\frac{1.65}{h}\right) = 0.5604\left(\frac{1.65}{0.176}\right) = 5.25$$

   $$\text{Inplane SIF} = C_1\left(\frac{0.9}{h^{2/3}}\right) = 0.5604\left(\frac{0.9}{0.176^{2/3}}\right) = 0.5604(2.86) = 1.6$$

   $$\text{Outplane SIF} = C_1\left(\frac{0.75}{h^{2/3}}\right) = 0.5604(2.4) = 1.345$$

By modeling flanges at the elbows, the lower SIF values can be advantageously used. However, the flexibility factor also has been reduced which is not desirable.

MITER BENDS

Miter bends shall be used, when more economical, for changes in direction on steel water piping, drain lines, and internal piping in pressure vessels in which space limitations prohibit the use of elbows. Miter bends in horizontal suction lines to centrifugal pumps should be a minimum of six pipe diameters from the suction flange. The equations to calculate stress intensification factors for miters are given in Table 4.3a. The miter bend can be either closely spaced or widely spaced as determined by using the following equations.

The miter is closely spaced if the miter space S is:

$$S < r_2(1 + \tan\theta) \qquad (4.4a)$$

$$R_1 = \text{Bend radius} = \frac{S \cot\theta}{2} \qquad (4.4b)$$

The miter is widely spaced if the miter space S is:

$$S \geq r_2(1 + \tan\theta) \qquad (4.4c)$$

$$R_1 = \frac{r_2(1 + \cot\theta)}{2} \qquad (4.4d)$$

where θ = miter angle, degrees
$\quad r_2$ = mean radius of the matching pipe, inches

(For maximum allowable internal pressure calculations, see Equations 2.8a, 2.8b, and 2.8c.)

The miter angle θ is equal to $11\frac{1}{4}$ for a five-piece (or four-weld) miter, sketch (d) in Table 4.4. θ is equal to 15 for a four-piece (or three-weld) miter, sketch (c), Table 4.4 and θ is equal to $22\frac{1}{2}$ for a three-piece (or two-weld) mitter, sketch (b) Table 4.4. Table 4.4 shows these miters and also gives miter space S.

Example

6. Calculate the SIF and flexibility factor k for an 8 in. four-piece miter. The plate thickness is 0.322 in.

For an 8 in. nominal pipe, r_2 = mean radius = 4.152 in. For a four-piece (three-weld) miter, θ = miter angle = 15°.

From Table 4.4, S = miter space = $6\frac{7}{16}$ in. Check for closely or widely spaced miter: $r_2(1 + \tan 15)$ = 5.26 in., which is less than the miter space from the table. Thus the given miter is a widely spaced miter.

TABLE 4.4 Miter Space Dimension for Miter Bends, in.

| (a) Single-weld miter | (b) Two-weld miter | (c) Three-weld miter | (d) Four-weld miter |

Pipe Diameter	Bend Radius R_1	Miter Space S		
		$\theta = 22\frac{1}{2}°$	$\theta = 15°$	$\theta = 11\frac{1}{4}°$
3	$4\frac{1}{2}$	$3\frac{3}{4}$	$2\frac{7}{16}$	$1\frac{13}{16}$
4	6	5	$3\frac{3}{16}$	$2\frac{3}{8}$
6	9	$7\frac{7}{16}$	$4\frac{13}{16}$	$3\frac{9}{16}$
8	12	$9\frac{15}{16}$	$6\frac{7}{16}$	$4\frac{3}{4}$
10	15	$12\frac{7}{16}$	$8\frac{1}{16}$	$5\frac{15}{16}$
12	18	$14\frac{15}{16}$	$9\frac{5}{8}$	$7\frac{3}{16}$
14	21	$17\frac{3}{8}$	$11\frac{1}{4}$	$8\frac{3}{8}$
16	24	$19\frac{7}{8}$	$12\frac{7}{8}$	$9\frac{9}{16}$
18	27	$22\frac{3}{8}$	$14\frac{7}{16}$	$10\frac{3}{4}$
20	30	$24\frac{7}{8}$	$16\frac{1}{16}$	$11\frac{15}{16}$
22	33	$27\frac{5}{16}$	$17\frac{11}{16}$	$13\frac{1}{8}$
24	36	$29\frac{13}{16}$	$19\frac{5}{16}$	$14\frac{5}{16}$
26	39	$32\frac{5}{16}$	$20\frac{7}{8}$	$15\frac{1}{2}$
28	42	$34\frac{3}{16}$	$22\frac{1}{2}$	$16\frac{11}{16}$
30	45	$37\frac{1}{4}$	$24\frac{1}{8}$	$17\frac{7}{8}$
32	48	$39\frac{3}{4}$	$25\frac{11}{16}$	$19\frac{1}{8}$
34	51	$42\frac{1}{4}$	$27\frac{5}{16}$	$20\frac{5}{16}$
36	54	$44\frac{3}{4}$	$28\frac{15}{16}$	$21\frac{1}{2}$
38	57	$47\frac{1}{4}$	$30\frac{9}{16}$	$22\frac{11}{16}$
40	60	$49\frac{11}{16}$	$32\frac{1}{8}$	$23\frac{7}{8}$
42	63	$52\frac{3}{16}$	$33\frac{3}{4}$	$25\frac{1}{16}$
48	72	$59\frac{5}{8}$	$38\frac{9}{16}$	$28\frac{5}{8}$

Using the equation from the code, Table 4.3a,

$$h = \text{flexibility characteristic} = \frac{1 + \cot\theta}{2}\frac{\bar{T}}{r_2} = \frac{1 + \cot 15}{2}\frac{0.322}{4.152} = 0.1835$$

$$k = \text{flexibility factor} = \frac{1.52}{(0.1835)^{5/6}} = 6.24$$

$$I_i = I_o = \text{SIF} = \frac{0.9}{(h)^{2/3}} = \frac{0.9}{(0.1835)^{2/3}} = \frac{0.9}{0.3229} = 2.78$$

Table 4.3a gives equations to calculate the flexibility factor and SIF for the following branch intersection types:

1. Welding tee
2. Reinforced fabricated tee with pad or saddle.
3. Unreinforced fabricated tee or stub-in
4. Extruded welding tee
5. Weld in contour insert (weldolet)
6. Branch welded on fitting

Branch intersections are sometimes identified by trade names or names given by a specific manufacturer. It is important to remember that SIF value should not be less than 1 (Note 1 of Table 4.3b).

When pad thickness T_r is greater than $1\frac{1}{2}$ times the pipe thickness $\bar{T}$, the equation to calculate h becomes

$$h = 4\frac{\bar{T}}{r_2} \qquad \text{(see Note 5, Table 4.3b)} \qquad (4.4e)$$

When this condition is reached h is no longer the function of pad thickness. That means credit cannot be obtained for a pad thickness portion that is greater than $1\frac{1}{2}$ times the pipe thickness.

SIF values for most branch intersection types are a function of run pipe dimensions and not branch pipe.

Example

7. Calculate SIF and k factor for an 8 in. diameter standard sch pipe with 4 in. branch if:

   (a) if intersection is an unreinforced fabricated tee
   (b) if pad thickness used is equal to pipe thickness
   (c) if pad thickness used is 0.57 in.

   The header wall thickness is $\bar{T} = 0.322$ in. The mean radius of the pipe is $r_2 = 4.152$ in.

(a) The unreinforced fabricated tee:

h = flexibility characteristic = $\bar{T}/r_2 = 0.322/4.152 = 0.0776$

k = flexibility factor = 1

i_o = outplane SIF = $0.9/h^{2/3} = 0.9/(0.0776)^{2/3} = 4.95$

i_i = inplane SIF = $\frac{3}{4}i_o + \frac{1}{4} = \frac{3}{4}(4.95) + \frac{1}{4} = 3.96$

(b) The reinforced fabricated tee:

$\bar{T}_r$ = pad thickness = 0.322 in., k = 1.0

$$h = \frac{(\bar{T} + \frac{1}{2}\bar{T}_r)^{5/2}}{\bar{T}^{3/2}r_2} = \frac{[0.322 + \frac{1}{2}(0.322)]^{5/2}}{(0.322)^{3/2}(4.152)} = \frac{0.16213}{0.1827(4.152)} = 0.213$$

$$i_o = \frac{0.9}{h^{2/3}} = \frac{0.9}{(0.213)^{2/3}} = \frac{0.9}{0.3566} = 2.52$$

$$i_i = \frac{3}{4}i_o + \frac{1}{4} = \frac{3}{4}(2.52) + \frac{1}{4} = 2.14$$

(c) The pad thickness = 0.57 in. See Note 5 of Table 4.3b. When $\bar{T}_r > 1\frac{1}{2}\bar{T}$, use $h = 4(\bar{T}/r_2)$.

$$1.5(0.375) = 0.5625$$

Given pad thickness $\bar{T}_r = 0.57$ in. $> 1.5(0.375)$.

$$\text{Thus } h = 4\frac{\bar{T}}{r_2} = 4\left(\frac{0.375}{4.152}\right) = 0.3613$$

$$i_o = \frac{0.9}{h^{2/3}} = \frac{0.9}{(0.3613)^{2/3}} = \frac{0.9}{0.507} = 1.18$$

$$i_i = \frac{3}{4}(i_o) + \frac{1}{4} = \frac{3}{4}(1.18) + \frac{1}{4} = 1.137$$

$$k = 1.0$$

EFFECT OF PRESSURE ON STRESS INTENSIFICATION AND FLEXIBILITY FACTORS

Some piping codes (references 2 and 3) give formulas for correcting flexibility factor and stress intensification factor (SIF) for elbows or bends. The effect of pressure on stress, forces, and moments by using corrected stress intensification factors and flexibility factors is discussed next.

When pressure effect is considered, SIF valves are lower, thus reducing the actual thermal stress. However, the anchor force increases because the flexibility at the bend has reduced. Pressure can affect significantly the magnitude of the flexibility factor and SIF in large diameter thin-wall elbows.

TABLE 4.5 Flexibility and Stress Intensification Factor for Bend

Description	Flexibility Factor k	Stress Intensification Factor		Flexibility Characteristic h	Sketch
		Outplane i_o	Inplane i_i		
Welding elbow or pipe bend	$\dfrac{1.65}{h}$	$\dfrac{0.75}{h^{2/3}}$	$\dfrac{0.9}{h^{2/3}}$	$\dfrac{\bar{T}R_1}{(r_2)^2}$	

Flexibility factors and SIF are very important constants in pipe stress calculations.

Table 4.5 gives equations for calculating flexibility characteristic h, flexibility factor k, outplane and inplane stress intensification factors (i_o and i_i) for elbow and bend.

The flexibility factor k in the table applies to bending in any plane. The flexibility factor k and stress intensification factor i should not be less than unity; factors for torsion equal unity. Both factors apply over the effective arc length (shown by heavy center lines in the sketch) for curved bends.

A single intensification factor equal to $0.9 h^{2/3}$ may be used for both i_i and i_o if desired.

The correction factor CFK for flexibility factor due to pressure on elbow or bend is given as Eq. 4.5a:

$$\text{CFK} = 1 + 6\left(\frac{P}{E_c}\right)\left(\frac{r_2}{\bar{T}}\right)^{7/3}\left(\frac{R_1}{r_2}\right)^{1/3} \qquad (4.5a)$$

The correction factor CFI for SIF is given as Eq. 4.5b:

$$\text{CFI} = 1 + 3.25\left(\frac{P}{E_c}\right)\left(\frac{r_2}{\bar{T}}\right)^{5/2}\left(\frac{R_1}{r_2}\right)^{2/3} \qquad (4.5b)$$

where $\bar{T}$ = nominal wall thickness of the fittings for elbows and miter bends, inches

r_2 = mean radius of matching pipe, inches

R_1 = bend radius of welding elbow or pipe bend, inches

P = gauge pressure, psi

E_c = cold modulus of elasticity, psi

Equations 4.5a and b for correction factors are given in Chemical Plant and Petroleum Refinery Piping (B31.3-1980) and Liquid Petroleum Transportation Piping Systems (B31.4-1979).

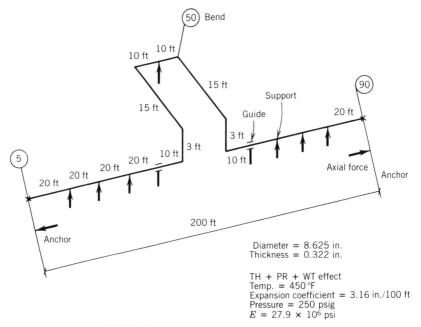

FIGURE 4.4   Symmetrical expansion loop.

To illustrate the effect of pressure on flexibility factor and stress intensification factors, an example problem using welding elbows (Node 50) is used as shown in Figure 4.4.

Example

8. The outside diameter of a pipe is 8.625 in.

$$r_2 = \frac{1}{2}\left[\frac{8.625+[8.625-2(0.322)]}{2}\right] = 4.1515 \text{ in.}$$

$\bar{T} = 0.322$ in.

$R_1 = 1.5(\text{nominal diameter}) = 1.5(8) = 12$ in.

$P = 250$ psig

$E_c = 27.9 \times 10^6$ psi

The flexibility characteristic is:

$$h = \frac{\bar{T}R_1}{(r_2)^2}$$

$$= \frac{0.322(12)}{(4.1515)^2} = 0.224$$

The outplane SIF:

$$i_o = \frac{0.75}{(h)^{2/3}} = \frac{0.75}{(0.224)^{2/3}} = 2.03$$

The inplane SIF:

$$i_i = \frac{0.9}{(h)^{2/3}} = 2.44$$

The flexibility factor is:

$$k = \frac{1.65}{h} = \frac{1.65}{0.224} = 7.366$$

The correction factor CFK for the flexibility factor (Eq. 4.5a)

$$= \left[1 + 6\left(\frac{P}{E_c}\right)\left(\frac{r_2}{T}\right)^{7/3}\left(\frac{R_1}{r_2}\right)^{1/3}\right]$$

$$= \left[1 + 6\left(\frac{250}{27.9 \times 10^6}\right)\left(\frac{4.1515}{0.322}\right)^{7/3}\left(\frac{12}{4.1515}\right)^{1/3}\right]$$

$$= 1.02967$$

The corrected flexibility factor is:

$$= \frac{7.366}{1.02967}$$

$$= 7.1537$$

Correction factor CFI for SIF (Eq. 4.5b)

$$= \left[1 + 3.25\left(\frac{P}{E_c}\right)\left(\frac{r_2}{T}\right)^{5/2}\left(\frac{R_1}{r_2}\right)^{2/3}\right]$$

$$= \left[1 + 3.25\left(\frac{250}{27.9 \times 10^6}\right)\left(\frac{4.1515}{0.322}\right)^{5/2}\left(\frac{12}{4.1515}\right)^{2/3}\right]$$

$$= 1.03527$$

$$\text{Corrected outplane SIF} = \frac{2.44}{1.03527} = 2.36$$

$$\text{Corrected inplane SIF} = \frac{2.03}{1.03527} = 1.96$$

TABLE 4.6 Effect of Pressure at Elbows (Node 50)
(pressure = 250 psi)

Pipe OD (inch)	Wall Thickness (inch)	Pressure Effect	Stress Intensification Factor		Stress		Force	
			i_i	i_o	Expansion Stress at 50 psi	% Change from no Pressure Effect	Axial Force at Anchors lb, F_z	% Change from no Pressure Effect
6.625	0.280	No	2.27	1.89	15,421	1.7%	487	0.61
	40 sch	Included	2.21	1.84	14,982	lower	490	larger
8.625	0.322	No	2.44	2.03	18,396	2.1%	1113	1.0%
	40 sch	Included	2.36	1.96	18,009	lower	1125	larger
10.75	0.250	No	3.40	2.84	22,053	5.4%	1285	4.6%
	20 sch	Included	3.04	2.53	20,845	lower	1345	larger
12.75	0.375	No	2.86	2.39	23,279	3.1%	3338	2.7%
	Std	Included	2.69	2.24	22,555	lower	3429	larger

Note that the flexibility factor and the stress intensification factor reduce when pressure effect on elbows is considered. Four different diameters were used ranging from 6 to 12 in. for result comparison.

Table 4.6 gives the results obtained. For example, with an 8.625 in. OD pipe, SIF factors are $i_i = 2.44$ and $i_o = 2.03$ when no pressure effect was considered, resulting in expansion stress of 18,396 psi at bend 50. When pressure effect was considered, for the same 8 in. nominal diameter pipe, $i_i = 2.36$, $i_o = 1.96$, and expansion stress of 18,009 psi at bend 50 were obtained.

The percent change in results due to the pressure effect for the 8 in. line is:

$$\text{percent change (lower) in expansion stress} = \frac{18,396 - 18,009}{18,396} \times 100$$

$$= 2.1\%$$

$$\text{percent change (larger) in axial force at anchor} = \frac{1125 - 1113}{1113} \times 100$$

$$= 1.0\%$$

For the same expansion loop with 8.625 in. OD pipe, the pressure range was changed. As Table 4.7 shows, the pressure effect becomes more significant with increase in pressure.

It is possible that including pressure effect on SIF and on flexibility factor could make the difference between the expansion stress valves obtained with and without pressure effect. This effect will be significant in the case of large diameter thin-wall elbows.

TABLE 4.7   Effect of Pressure at Elbows (Node 50)
OD = 8.625 in., thickness = 0.322 in., temperature = 450°F

Pressure (psig)	Stress Intensification Factor		Expansion Stress (psi)	% Change (lower)	Axial Force at Anchor (lb)	% Change (larger)
	i_i	i_o				
No pressure	2.44	2.03	18,396	—	1113	—
250	2.36	1.96	18,009	2.1	1125	1
350	2.32	1.94	17,975	2.2	1137	2
450	2.29	1.91	17,942	2.5	1148	3
550	2.26	1.89	17,910	2.6	1159	4
650	2.23	1.86	17,878	2.8	1170	5

As stated before only B31.3 and B31.4 piping codes have allowed the use of using Eq. 4.5b to include the effect of pressure on SIF. Basic work on this area and the formulation for the equations is found in reference 7 and this information was used to reduce the stress in piping in real case analysis. Two large diameter (65.74 in.) long (6500 ft) steam lines were built to supply saturated steam at 400°F to heavy water plants at Ontario Hydro's Bruce Nuclear Power Development (reference 8). In preliminary analysis, the equations for flexibility and stress intensification factors given in power code B31.1 were used (reference 1). In further analysis Eqs. 4.5a and b were used and the piping was qualified.

STRESSES IN A PIPING SYSTEM

The equation for expansion stress S_E is given by Eq. 4.6. The equation for resultant bending stress S_b is given by Eq. 4.7. For branch connections, the resultant bending stress equation requires attention because the section modulus value Z used for header and branch is slightly different. Equations 4.8 and 4.9 show this difference. The calculated value of expansion stress S_E needs to be lower than expansion stress range S_A, earlier defined by Eq. 4.1.

The stresses in a piping system is generally low for smaller temperature variation, smaller diameter, smaller expansion coefficient, lower modulus of elasticity, and the longer the length of the pipe in a direction perpendicular to direction of expansion.

The pipe wall thickness has no significant effect on bending stress due to thermal expansion but it affects the end forces and moments in direct ratio. So overstress cannot be remedied by adding thickness; on the contrary, this tends to make matters worse by increasing forces and moments.

Flexibility Stresses*

1. Bending and torsional stresses shall be computed using the as-installed modulus of elasticity E_a and then combined in accordance with Eq. 4.6 to determine the computed displacement stress range S_E, which shall not exceed the allowable stress range S_A:

$$S_E = \sqrt{S_b^2 + 4S_t^2} \qquad (4.6)$$

where S_b = resultant bending stress, psi
  $S_t = M_t/2Z$ = torsional stress, psi
  M_t = torsional moment, in.-lb
  Z = section modulus of pipe, in.^{3}

---

* From ANSI/ASME B31.3-1980, Section 319.4.4.

The thermal expansion stress (Eq. 4.6) is based on the maximum shear theory.

2. The resultant bending stresses S_b to be used in Eq. 4.6 for elbows and miter bends shall be calculated in accordance with Eq. 4.7, with moments as shown in Figure 4.2.

$$S_b = \text{resultant bending stress, psi}$$

$$= \frac{\sqrt{(i_i M_i)^2 + (i_o M_o)^2}}{Z} \tag{4.7}$$

where i_i = inplane stress intensification factor from Table 4.3a
$\quad\quad i_o$ = outplane stress intensification factor from Table 4.3a
$\quad\quad M_i$ = inplane bending moment, in. lb
$\quad\quad M_o$ = outplane bending moment, in. lb
$\quad\quad Z$ = sectional modulus of pipe, in.^{3}

3. The resultant bending stresses S_b to be used in Eq. 4.6 for branch connections shall be calculated in accordance with Eqs. 4.8 and 4.9 with moments as shown in Figure 4.3. For header (legs 1 and 2):

$$S_b = \frac{\sqrt{(i_i M_i)^2 + (i_o M_o)^2}}{Z} \tag{4.8}$$

For branch (leg 3):

$$S_b = \frac{\sqrt{(i_i M_i)^2 + (i_o M_o)^2}}{Z_e} \tag{4.9}$$

where S_b = resultant bending stress, psi
$\quad\quad Z_e$ = effective section modulus for branch of tee, in.^{3} = $\pi r_m^2 t_s$
$\quad\quad r_m$ = mean branch cross-sectional radius, inches
$\quad\quad t_s$ = effective branch wall thickness, inches (lesser of t_h and $(i_o)(t_b)$)
$\quad\quad t_h$ = thickness of pipe matching run of tee or header exclusive of reinforcing elements, inches
$\quad\quad t_b$ = thickness of pipe matching branch, inches
$\quad\quad i_o$ = outplane stress intensification factor
$\quad\quad i_i$ = inplane stress intensification factor

4. Allowable stress range S_A and permissible additive stresses shall be computed in accordance with Eq. 4.1 and Eq. 4.2.

Example

9. Calculate torsional stress, bending stress, and expansion stress at the intersection of 4 in. sch 80 header and 3 in. sch 40 branch (Fig. 4.5).

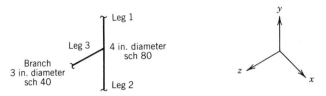

FIGURE 4.5   Stresses at run and branches.

Moments acting at the intersection is given below:

		M_x ft-lb	M_y	M_z
Header	leg 1	−550	2322	800
Header	leg 2	425	1821	−890
Branch	leg 3	−180	−920	682

Solution: For header, y moment is torsion; x moment is inplane; and z moment is outplane moment.

For branch, y moment is outplane; x moment is inplane; and z moment is torsional moment.

Assume unreinforced fabricated tee (stub-in), that gives the highest number for SIF.

$$\bar{T} = 0.337 \text{ in. (see Appendix Table A4 for 4 in. sch 80)}$$

$$r_2 = \frac{1}{2}(4.5 - 0.337) = 2.082 \text{ in.}$$

$$Z = 4.27 \text{ in.}^3$$

$$h = \frac{\bar{T}}{r_2} = \frac{0.337}{2.082} = 0.162$$

$$i_o = \frac{0.9}{h^{2/3}} = \frac{0.9}{(0.162)^{2/3}} = 3.03$$

$$i_i = 0.75(3.03) + 0.25 = 2.52$$

For Header Leg 1:   The resultant bending stress is:

$$S_b = \frac{\sqrt{(i_i M_i)^2 + (i_o M_o)^2}}{Z}$$

$$S_b = \frac{\sqrt{[2.52 \times (-550)12]^2 + [3.03(800)12]^2}}{4.27} \qquad (4.10)$$

$$= \frac{41{,}136}{4.27} = 9638 \text{ psi}$$

The torsional stress is:

$$S_t = \frac{M_t}{2Z} = \frac{M_y}{2(Z)} = \frac{2322 \times 12}{2(4.27)} = 3262 \text{ psi}$$

The computed displacement stress range is:

$$S_E = \sqrt{S_b^2 + 4S_t^2}$$
$$= \sqrt{9638^2 + 4(3262)^2} = 13,630 \text{ psi}$$

For Header Leg 2:

$$S_b = \frac{\sqrt{(2.52 \times 425 \times 12)^2 + [3.03(-890)12]^2}}{4.27} = \frac{45,764}{4.27}$$
$$= 10,717 \text{ psi}$$

$$S_t = \frac{1821 \times 12}{2(4.27)} = 2558$$

$$S_E = \sqrt{10,717^2 + 4(2558)^2} = 11,876 \text{ psi}$$

For Branch Leg 3:

$$\bar{T} = 0.216 \text{ in.} \qquad Z = 1.724 \text{ in.}^3 \quad \text{(Appendix Table A4)}$$

Z_e = effective section modulus for branch

$$= \pi r_m^2 t_s$$

$$r_m = \tfrac{1}{2}(3.5 - 0.216) = 1.642 \text{ in.}$$

t_s = lesser of t_h or $(i_o t_b)$

$$t_h = 0.337 \text{ in.} \quad (\bar{T} \text{ and } t_h \text{ mean the same thing})$$

$$i_o(t_b) = 3.03(0.216) = 0.654 \text{ in.}$$

$$t_s = 0.337 \text{ in.}$$

$$Z_e = \pi(1.642)^2(0.337) = 2.85 \text{ in.}^3$$

$$S_b = \frac{\sqrt{[2.523(-180)12]^2 + [3.03(-920)12]^2}}{2.85} = \frac{33,892}{2.85}$$

$$= 11,892 \text{ psi} \qquad\qquad (4.11)$$

$$S_t = \frac{M_t}{2(Z)} = \frac{682(12)}{2(1.724)} = 2373 \text{ psi}$$

$$S_E = \sqrt{11,892^2 + 4(2373)^2} = 12,804 \text{ psi}$$

Note the same SIF is used for the header and branch. In bending stress calculation (Eq. 4.11) for branch, the section modulus value is modified.

COLD SPRING

A piping system may be cold sprung or prestressed to reduce anchor force and moments caused by thermal expansion. Cold spring may be cut short for hot piping or cut long for cold (cryogenic) piping. The cut short is accomplished by shortening the overall length of pipe by desired amount but not exceeding the calculated expansion. Cut long is done by inserting a length (making the length of pipe longer). The amount of cold spring (CS) is expressed as a percentage or fraction of thermal expansion.

Credit for cold spring is *not* allowed for stress calculations. Different codes state the same meaning by slightly different wording. The following is from Nuclear Code, Class 2 piping NC-3673-3 ASME Section III:

> **NC-3673.3 Cold Springing.**   No credit for cold spring is allowed with regard to stresses. In calculating end thrusts and moments acting on equipment, the actual reactions at any one time, rather than their range, shall be used. Credit for cold springing is allowed in the calculations of thrusts and moments, provided the method of obtaining the designed cold spring is specified and used.

Figure 4.6 shows the position of the pipe before and after cold spring (cut short in this case). The length of pipe is 85 ft in x direction and grows 1.54 in. at the temperature of 300°F. The percentage cold spring desired is 50%. The amount of length to be cut short is equal to the product of percentage cold spring and actual expansion; here it is 0.77 in. For practical reasons, to achieve the same at the construction site cold spring of $\frac{3}{4}$in. is used. As can be seen the pipe is pulled back during installation. This is done by physical force using equipment such as a tractor. When the pipe gets hot, it crosses the neutral position and grows toward the other side.

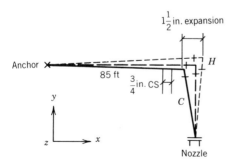

C = installed position (cold)
H = position after expansion (hot)

FIGURE 4.6 Piping in initial (cold) and final (hot) position under cold spring.

The following difficulties have been faced with respect to cold spring:

1. By mistake some calculate the stress also with cold spring, which is wrong. First the system should pass for stress calculations without considering cold spring. Only in the next calculation can cold spring be considered for reducing loads at equipments.
2. Construction crews sometimes overlook the need for cold spring and thus do not use the cold spring. The amount of force needed to pull a large line to the initial position for welding is huge.
3. The cold reaction at full cold spring needs to be calculated and made sure that the equipment is able to withstand this additional load due to total cold spring at cold condition.
4. The deflections at a cold spring location still remain the same, because the cold spring only relocates the pipe weld point and does not reduce the actual expansion. This actual deflection is important in the spring design. If attention is not paid, the spring may be undersized for deflection.
5. Cold spring needs to be specified at weld points to save cost of additional welding.

Maximum Reactions for Simple Systems*

For two-anchor systems without intermediate restraints, the maximum instantaneous values of reaction forces and moments may be estimated from B31.3 Eqs. 4.12 and 4.13a.

(*a*) *For Extreme Displacement Conditions* R_m: The temperature for this computation is the maximum or minimum metal temperature, whichever produces the larger reaction:

$$R_m = R\left(1 - \frac{2C}{3}\right)\frac{E_m}{E_a} \tag{4.12}$$

C = cold-spring factor varying from zero for no cold spring to 1.0 for 100% cold spring. (The factor 2/3 is based on experience that shows that specified cold spring cannot be fully assured, even with elaborate precautions.) Usually C of 0.5 is recommended.

E_a = modulus of elasticity at installation temperature

E_m = modulus of elasticity at maximum or minimum metal temperature

R = range of reaction forces or moments (derived from flexibility analysis) corresponding to the full displacement stress range and based on E_a

R_m = estimated instantaneous maximum reaction force or moment at maximum or minimum metal temperature

---

* From ANSI B31.3, subsection 319.5.1 and 319.5.2.

(*b*)   *For Original Condition* R_a: The temperature for this computation is the expected temperature at which the piping is to be assembled.

$$R_a = CR \quad \text{or} \quad C_1 R, \text{ whichever is greater} \tag{4.13a}$$

where nomenclature is as before and:

$$C_1 = 1 - \frac{S_h E_a}{S_E E_m} \tag{4.13b}$$

C_1 = estimated self-spring or relaxation factor; use zero if value of C_1 is
   negative.
R_a = estimated instantaneous reaction force or moment at installation temperature
S_E = computed displacement stress range (from Eq. 4.6)
S_h = hot stress, psi

Maximum Reactions for Complex Systems

For multianchor systems and for two-anchor systems with intermediate restraints, Eqs. 4.12, 4.13a and b are not applicable. Each case must be studied to estimate location, nature, and extent of local overstrain and its effect on stress distribution and reactions.

   If a piping system is designed with different percentages of cold spring in various directions, these equations are not applicable. In this case, the piping system shall be analyzed by a comprehensive method. The calculated hot reactions shall be based on theoretical cold springs in all directions not greater than two thirds of the cold springs as specified or measured.

Example

10.   Calculate cold and hot reaction moments at nozzle (Fig. 4.7) after 55% cold spring if moment without cold spring was 2500 ft-lb from piping analysis.

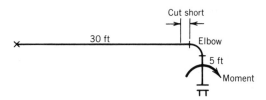

FIGURE 4.7   Moment calculation under cold spring.

Piping material is stainless steel A312 TP 304. The temperature is 900°F.

To calculate *hot reaction*, use Eq. 4.12:

$$R_m = R\left(1 - \frac{2C}{3}\right)\frac{E_m}{E_a}$$

where R = moment before cold spring = 2500 ft-lb
    $C = 0.55$
    E_m = hot modulus = 23.4×10^6 psi, at 900°F for stainless steel (Appendix Table A2)
    E_a = cold modulus = 28.3×10^6 psi

$$R_m = 2500\left(1 - \frac{2(0.55)}{3}\right)\frac{23.4 \times 10^6}{28.3 \times 10^6}$$

$$= 2500(0.37)(0.8269)$$
$$= 758 \text{ ft-lb.}$$

To calculate *cold reaction* use Eq. 4.13a:

$$R_a = CR \quad \text{or} \quad C_1 R, \text{ whichever is greater}$$

where C_1 = relaxation factor

$$= 1 - \frac{S_h}{S_E}\frac{E_a}{E_m}$$

Because there was not enough information to calculate computed expansion stress range S_E, factor C_1 could not be calculated.
    Cold reaction, $R_a = CR$
$$= 0.55(2500)$$
$$= 1375 \text{ ft-lb}$$

It is important that the equipment nozzle should withstand not only 758 ft-lb in an operating condition, but also 1375 ft-lb in a cold condition.

EXERCISES

1. Find cold stress and hot stress for a carbon steel seamless pipe at −30°F, 675°F, 1125°F:
(a) material is A53 Grade A; (b) material is API 5 L Grade B.

2. A rotating equipment nozzle can only allow a force of 800 lb during operation (Fig. 4.8). The carbon steel pipe will have an operating temperature 500°F and a calculated force of 3520 lb. A cold spring can be

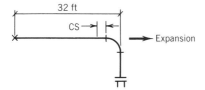

FIGURE 4.8 Cold spring example.

used to reduce the force. What is the minimum percentage of cold spring should be used?

3. Calculate the reaction force in a stainless steel piping system with 65% cold spring. Operating temperature is 800°F and force without cold spring is 982 lb.

4. At what condition can cold spring be used? List difficulties encountered with cold spring in theory and practice.

5. (a) Calculate allowable stress range for A53 Grade A material pipe at 682°F.
   (b) Calculate longitudinal stress in a 12 ft standard weight welding elbow when:
   Inplane bending moment   = 423 ft-lb
   Outplane bending moment = 325 ft-lb
   Axial force   = 628 lb
   Material is A53 Grade A and temperature is 682°F.

6. Calculate SIF and flexibility factor:
   (a) 6 in. long radius standard thickness
   (b) Calculate corrected SIF and k if the elbow is two ends flanged.
   (c) Miter bend with $\theta = 15°$ and 12 in. diameter with thickness 0.25 in.

7. (a) Calculate thermal expansion force in the piping shown in Figure 4.9. The pipe is 16 in. sch 80 A53 Grade B pipe at 600°F.
   (b) If the distance between anchors is increased to 300 ft, what will be the force?

8. For a seamless pipe, A53 Grade B, the allowable stresses at 70°F and 600°F are 20,000 psi and 17,300 psi, respectively according to ANSI B31.3 code. For an actual piping system at 600°F, the computed piping stresses at certain locations are as follows:
   (a) Longitudinal stress due to pressure weight and other sustained loading is 9800 psi.
   (b) Computed displacement stress range is 33,475 psi.
   (c) Stresses due to wind load is 5822 psi.
   Does this piping system meet the stress criteria for ANSI B31.3 code?

FIGURE 4.9 Axial force in restrained piping.

9. Calculate the thermal expansion stress for the branch and the header according to ANSI B31.3 code for the loading (given in Example 9 and in Fig. 4.5) at the branch intersection: the branch and the header are 12 in. standard wall and 8 in. sch 40 wall.

REFERENCES

1. ANSI B31.1-1980. Power Piping Code.
2. ANSI B31.3-1980. Chemical Plant and Petroleum Refinery Piping Code.
3. ANSI B31.4-1974. Liquid Transportation Piping Code.
4. ANSI B31.8. DOT. Gas Transmission Transportation Piping Code.
5. ASME. Section III, Nuclear Components Code.
6. Markl, Arc "Fatigue Tests of Piping Components," *Trans. ASME*, Vol. 74(3), pp. 287–303 (April 1952).
7. Rodabough, E. C. "Effect of Internal Pressure on Flexibility and SIF on Curved Pipe," *Journal of Applied Mechanics*, Vol. 24; *Trans ASME*, Vol. 79 (May 1957).
8. Machacek, S. "Design and Operation of a Large Diameter Steam Line at Ontario Hydro's Bruce Nuclear Power Development," ASME, 78-PVP-86.

CHAPTER FIVE

EXPANSION LOOPS AND EXPANSION JOINTS

As described earlier in Chapter 1, two devices used to improve the flexibility of piping are expansion loops and expansion joints. This chapter will deal with these two topics in more detail.

EXPANSION LOOPS

Loops provide the necessary leg of piping in a perpendicular direction to absorb the thermal expansion. They are safer when compared with expansion joints but take more space. Expansion loops may be symmetrical (Fig. 5.1) or nonsymmetrical (Fig. 5.2). Symmetrical loops are advantageous to use because leg H (Fig. 5.1) is used efficiently to absorb an equal amount of expansion from both directions. The bend length L_2 is given by:

$$L_2 = W + 2(H) \tag{5.1}$$

Sometimes nonsymmetrical loops are used to utilize the existing support steel or to locate the loop at road crossings. Vertical direction supports are provided to support the gravity weight at the calculated span as discussed in Chapter 3. Horizontal loops (bend length either flat or horizontal) would need a few more supports when compared with vertical loops in the bend length portion, as shown by supports $S_1 S_1$ in Figure 5.3. The optimum ratio of height per width can be estimated and used.

When several piping loops are laid side by side on a pipe rack, the size of the loop including the ratio height per width may be modified to lay the loops one inside the other as shown in Figure 5.3. But the final size of each loop (bend length) must be larger than the calculated bend length.

Hotter and larger lines are placed outside as outer loops because the longer height H is needed. Smaller lines with lower temperatures are placed as inside

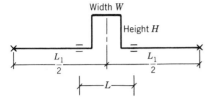

FIGURE 5.1 Symmetrical loop.

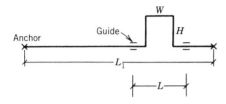

FIGURE 5.2 Nonsymmetrical loop.

loops. Because this loop arrangement may change the entire pipe rack layout, it is advisable to estimate the loop's sizes with simplified calculations or nomographs (Fig. 5.12) at early stages of the project. Guides on both sides of the loop, shown as G_1 and G_2 in Figure 5.3, are important for proper functioning of loops because guides direct the expansion into the bend E_1 along the axis of the pipe, which avoids shifting the lines sideways. A practical problem often encountered is interference when sufficient gap was not provided for in the design. The gap, after considering insulation on both lines, should be larger than the differential expansion at elbows E_1 and E_2 as shown in Figure 5.4. To avoid interference, gap $> (\Delta x2 - \Delta x1)$, where $\Delta x2$ and $\Delta x1$ are expansions occurring in the same direction at the same time.

See Figure 5.4 showing gap requirement and also the considering of insulation.

Caution should be exercised in calculating the differential expansion if the inner loop is not operating and is at 70°F. In this case, the actual gap is less as shown in Figure 5.5. Figure 5.6a shows that without guides the loop expansion is not directed properly. Figure 5.6b shows that the pipe occupies (known as snakes) space needed for future piping layout. This figure demonstrates the necessity for guides.

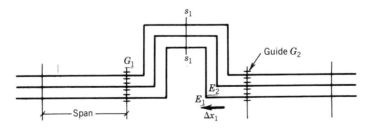

FIGURE 5.3 Layout in plan of many horizontal loops.

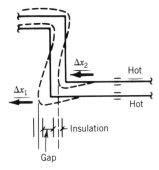

FIGURE 5.4 Gap requirement with both lines hot.

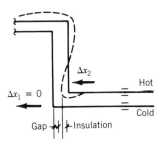

FIGURE 5.5 Gap requirement, inner or smaller loop is not operating.

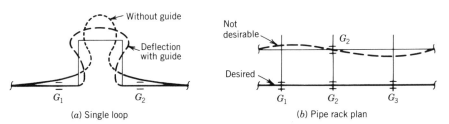

FIGURE 5.6 (a) Need for guides to control the direction of deflection. (b) Pipe rack with and without enough guides.

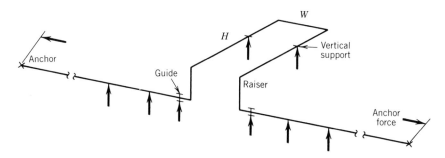

FIGURE 5.7 Three-dimensional horizontal loop.

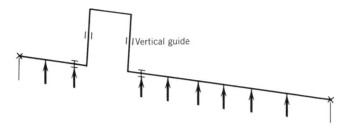

FIGURE 5.8   Vertical loops at road crossings.

Three-dimensional loops (Fig. 5.7) are widely used because this arrangement does not block the routing of low temperature lines under the loop. The usual raiser height is about 3 ft. The loop bend length L_2 is also taken here as $L_2 = W + 2H$ without giving credit for the two raisers.

Vertical loops are placed at road crossings and sometimes are nonsymmetrically located due to the location of the road. Vertical guides may be necessary to keep the line vertical as shown in Figure 5.8.

Stresses and Loads in Loops

Calculation of stress and loads in loops by the M. W. Kellogg method (reference 3 in Chapter 1) follows.

Sample Calculation

Given a loop of 20 in. OD $\times \frac{1}{2}$ in. thick ASTM A-135, Grade A pipe. K_1L is 20 ft. Guides are located 10 ft on either side of the loop, so that $L = 40$ ft. The distance between anchors A' and B' is 100 ft (Fig. 5.9). The line temperature is 425°F and is used for oil piping. Find

(a)   The required height of K_2L and
(b)   The forces acting at points A' and B' and the moments acting at points A and B.

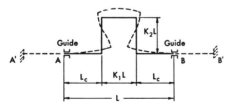

Symmetrical Expansion Loop Subjected
to Thermal Expansion

FIGURE 5.9   Stress and loads in a symmetrical loop.

(a) The unit linear thermal expansion for carbon steel at $425°F =$ 0.030 in./ft; $\Delta = 100 \times 0.03 = 3$ in. $S_A = 19,890$ psi (ignoring code permission to exclude longitudinal joint efficiency):

$$\frac{L^2 S_A}{10^7 D \Delta} = 0.0531$$

Enter in Figure 5.10 with 0.0531

Read over to the curve representing $K_1 = 0.5$ and down to the value of K_2, which is 0.32. $K_2 L$ is therefore $40 \times 0.32 = 12.80$ ft.

$$F_{xA'} = -F_{xB'} = -10^6 A_1 \left(\frac{I \Delta}{L^3} \right) \tag{5.2}$$

$$M_{zA} = -M_{zB} = 10^5 A_2 \left(\frac{I \Delta}{L^2} \right) \tag{5.3}$$

(b) The moment of inertia for 20 in. OD $\times \frac{1}{2}$ in. thick pipe $= 1457$ in.[4]

$$\frac{I \Delta}{L^3} = \frac{1457 \times 3}{64,000} = 0.0683$$

$$\frac{I \Delta}{L^2} = \frac{1457 \times 3}{1600} = 2.73$$

Enter in Figure 5.11 with $K_1 = 0.5$ and $K_2 = 0.32$.
Read

$A_1 = 0.55$

$A_2 = 0.86$

$F_{xA} = F_{xB} = -68,300 \times 0.55 = 37,600$ lb   (using Eq. 5.2)

$M_{zA} = -M_{zB} = +273,000 \times 0.86 = 235,000$ ft-lb   (using Eq. 5.3)

The nomograph presented as Figure 5.12 may be used to estimate the size of the expansion loop. Out of the four arrangements of single-plane expansion loops shown, type A is popular due to the ease in fabrication using standard elbows and straight pipe lengths. Other arrangements require pipe bending to a specific configuration.

In arriving at the nomograph, the following were assumed:
The formula used is the guided cantilever formula given by Eq. 1.3.

Leg required:

$$L_2 = \sqrt{\frac{3 E D \Delta}{144 S_A}} \tag{1.3}$$

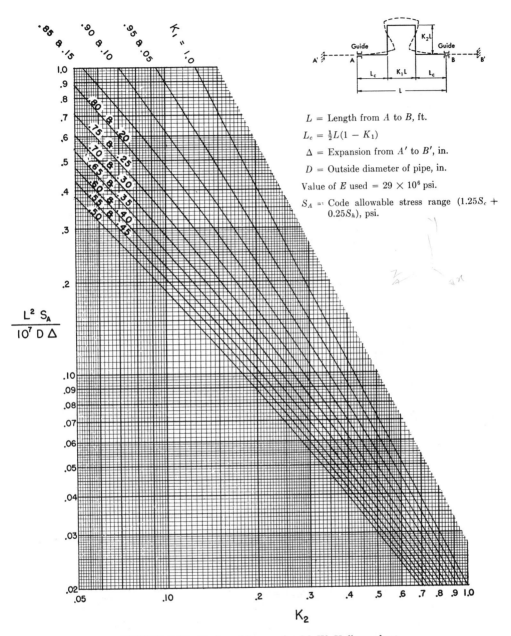

L = Length from A to B, ft.

$L_c = \frac{1}{2}L(1 - K_1)$

Δ = Expansion from A' to B', in.

D = Outside diameter of pipe, in.

Value of E used = 29×10^6 psi.

S_A = Code allowable stress range $(1.25S_c + 0.25S_h)$, psi.

FIGURE 5.10 Design of loops using M. W. Kellogg chart.

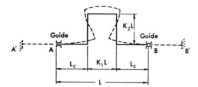

$$F_{zA'} = -F_{zB'} = -10^6 A_1 I\Delta/L^3$$

$$M_{zA} = -M_{zB} = 10^5 A_2 I\Delta/L^2$$

I = Moment of inertia of pipe, in.[4].
Δ = Expansion from A' to B', in.
Value of E used = 29×10^6 psi.

F = Force, lb.
M = Moment, ft-lb.
First subscript denotes *direction*.
Second subscript denotes *location*.
Signs are those of forces or moments *acting* on anchors.

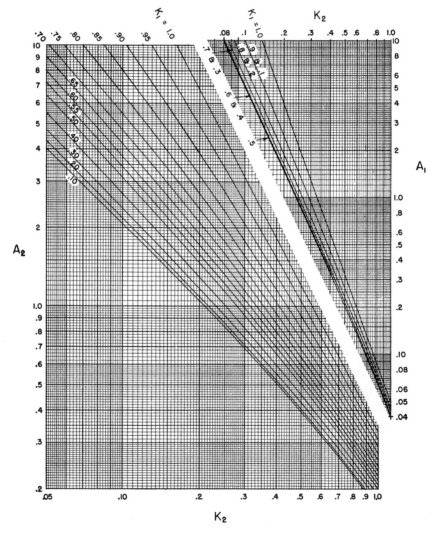

FIGURE 5.11 Moments and forces in a loop using M. W. Kellogg chart.

88

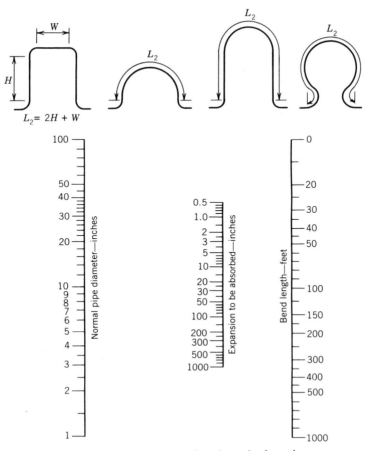

FIGURE 5.12 Nomograph to determine loop size.

where $S_A = 20,000$ psi

$E = 29 \times 10^6$ psi

Δ = expansion to be absorbed by the loop, inches

D = nominal diameter, inches (Note that Eq. 1.3 uses outside diameter)

L = Distance between guides, ft

L_1 = Distance between anchors, ft

L_2 = Bend length required to absorb expansion, ft

Example

1. Find the size of the loop to absorb expansion in 200 ft of 12 in. carbon steel pipe at 400°F. Assume height to width ratio.

$$\text{Total expansion} = 200(0.027) = 5.4 \text{ in.}$$

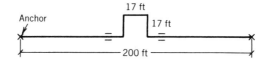

FIGURE 5.13   Estimated loop size using nomograph.

Using the nomograph and assuming a straight line starting from a 12 in. diameter and through a 5.4 in. expansion, read bend length L_2 as 50 ft. Assume $H = W$, then $L_2 = 2H + W = 50$ ft. Thus $H = W = 17$ ft, making $L_2 = 51$ ft.

By calculation

$$L_2 = \sqrt{\frac{3(29 \times 10^6)12(5.4)}{144(20,000)}} = 44 \text{ ft}$$

The estimated loop size is given in Figure 5.13.

2. Using the Kellogg method, calculate stress, force, and moment in the expansion loop shown in Figure 5.14. The pipe diameter is 6 in. sch 40, the temperature is 450°F and has carbon steel piping. Use Figures 5.10 and 5.11 to arrive at the solution. This problem is the same for which results were presented in Chapter 1, Table 1.3. Here the problem is solved step by step.

The expansion coefficient for carbon steel at 450°F = 0.0316 in./ft (Appendix A1)

$$\text{constant } K_1 = \frac{W}{L} = \frac{\text{width}}{\text{guide distance}} = \frac{20}{40} = 0.5$$

$$K_2 = \frac{H}{L} = \frac{\text{height}}{\text{guide distance}} = \frac{20}{40} = 0.5$$

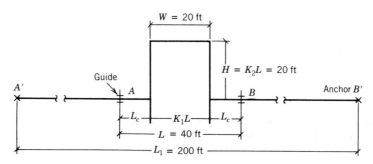

FIGURE 5.14   Stress and loads calculation using Kellogg method.

For $K_1 = 0.5$ and $K_2 = 0.5$, from Figure 5.10, read $L^2 S_A/10^7 \, DA$ as 0.03

$$L_c = \tfrac{1}{2}(L)(1 - K_1)$$
$$= \tfrac{1}{2}(40)(1 - 0.5) = 10 \, \text{ft}$$

Deflection $\Delta = 200(0.0316) = 6.32$ in.

$$OD = 6.625 \, \text{in.}$$

$$\text{Stress} = S = \frac{0.03(10^7)D\,\Delta}{L^2}$$

$$= \frac{0.03(10^7)(6.625)(6.32)}{40^2}$$

$$= 7850 \, \text{psi}$$

Moments and Forces

Using Figure 5.11, read:

$$A_1 = 0.21 \qquad A_2 = 0.5$$
$$\text{where } K_1 = 0.5 \qquad K_2 = 0.5$$
$$I = 28.14 \, \text{in.}^4$$

Axial force at anchor (Eq. 5.2):

$$F_{yA'} = -F_{xB'}$$

$$= -\frac{10^6 A_1 \, \Delta}{L^3}$$

$$= -\frac{10^6(0.21)(6.32)}{64,000}$$

$$= 584 \, \text{lb}$$

Moment at guides (Eq. 5.3):

$$M_{zA} = -M_{zB}$$

$$= \frac{10^5 A_2 I \, \Delta}{L^2}$$

$$= \frac{10^5(0.5)(28.14)(6.32)}{1600}$$

$$= 5558 \, \text{ft-lb}$$

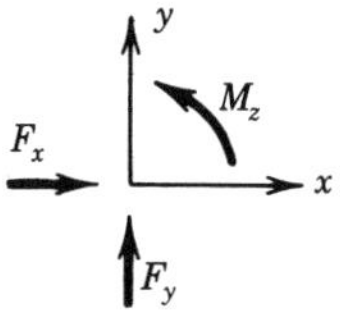

FIGURE 5.15   Coordinate used.

Note: First subscript denotes direction; second subscript denotes location. Signs are those of forces and moments acting on anchors (see Fig. 5.15).

EXPANSION JOINTS

In 1984 expansion joints were allowed in nuclear piping design except for the ASME Section III, Nuclear Class 1 Code. Subsection NB-3671.2 states that expansion joints are not allowed in Class 1 NB nuclear components. Past accidents with expansion joint installations are of concern from a safety point of view. Expansion joints are used to absorb axial compression or extension lateral offset and angular rotation. As per standards of the Expansion Joint Manufactures (reference 1), torsional rotation should be avoided on the bellows because torque produces high stress levels in bellows.

Expansion joints can be broadly classified as sliding and flexible. There is a relative motion of adjacent parts in the case of slipping joints. Slip joints, swivel joints, and ball joints are grouped under sliding joints. Dresser coupling and Victaulic couplings are a few trade names of joints of this type. Sliding joints are also known as packed joints because packing to contain internal pressure without leakage is necessary. Flexible expansion joints may be further divided into bellows joints, metal hose, and corrugated pipe (references 2 and 3).

The following are terms used in the design and specification of expansion joints (see Fig. 5.16 for symbols used for some of the terms):

Main Anchor: A main anchor must be designed to withstand the forces and moments imposed upon it by each of the pipe sections to which the anchor is attached. In the case of a pipe section containing an expansion joint, forces and moments will consist of the thrust due to pressure (Eq. 5.4), the force required to deflect the expansion joint (Eq. 5.5), and the frictional forces due to pipe alignment guides and supports. When a main anchor is installed at the change of direction of flow, the effect at the elbow of the centrifugal thrust due to flow (Eq. 5.6) must also be considered.

Intermediate Anchor: An intermediate anchor divides a pipe line into individual expanding pipe sections, each of which is made flexible through the use of one or more expansion joints.

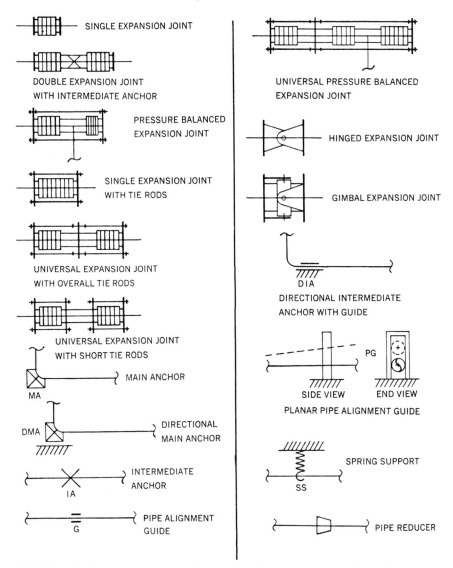

FIGURE 5.16 Types of expansion joints (standards of the Expansion Joint Manufactures Association).

Directional Anchor: A directional anchor or sliding anchor is one designed to absorb loading in one direction while permitting motion in another direction.

Bellows: The flexible element of an expansion joint consists of one or more corrugations and the tangents, if any.

Bellows Material: A list of metal bellows materials is given below:

Bellows material	Temperature Range °F (specified by ASME Section VIII)
304 stainless steel	−300 to 750
316 stainless steel	−300 to 750
321 stainless steel	−300 to 1500
347 stainless steel	−300 to 1400
Nickel 200	−300 to 600
Monel 400	−300 to 900
Inconel 600	−250 to 1200
Inconel 625	−250 to 1200
Incoloy 800	−250 to 1500
Incoloy 825	−250 to 800

Squirm in a Bellows Expansion Joint: A term employed to denote the occurrence of instability due to internal pressure and is predominately associated with joints of 20 in. diameter or smaller.

Flexibility of an Expansion Joint: This can be increased by thinner bellows (must still be able to withstand the pressure), increase in number of bellows, and by multiple bellows.

External Cover: A cover used to protect the exterior of the bellows from foreign objects, especially when the joint is buried underground.
   Internal liner or sleeve is used for the following:

1.   where it is necessary to minimize frictional loss
2.   where flow velocities are high (for steam lines when velocity exceeds 1000 ft/min/in. of diameter in lines upto 6-in. size)
3.   when abrasive materials are present
4.   when there is reverse or turbulent flow
5.   for all high temperature applications
6.   for all copper elbows

   When lateral deflection or rotation is present, the liner must be sufficiently smaller in diameter to provide the necessary clearance.

Tie Rods: These are rods or bar devices for the purpose of restraining the expansion joint from the thrust due to internal pressure. The number and size of the rods depend upon the magnitude of thrust force. Tie rods may also act as deflection limit rods.

Equalizing or Reinforcing Rings: These help to reinforce the elbows against internal pressure and help to maintain the desired shape of the elbows.

Guides: Guides are important parts of expansion joint performance.

TYPES OF EXPANSION JOINTS (see Fig. 5.16)

SINGLE EXPANSION JOINT: The simplest form of expansion joint, of single bellows construction, designed to absorb all of the movement of the pipe section in which it is installed.

DOUBLE EXPANSION JOINT: A double expansion joint consists of two bellows joined by a common connector which is anchored to some rigid part of the installation by means of an anchor base. The anchor base may be attached to the commom connector either at installation or at time of manufacture. Each bellows acts as a single expansion joint and absorbs the movement of the pipe section in which it is installed independently of the other bellows. Double expansion joints should not be confused with universal expansion joints.

INTERNALLY GUIDED EXPANSION JOINT: An internally guided expansion joint is designed to provide axial guiding within the expansion joint by incorporating a heavy telescoping internal guide sleeve, with or without the use of bearing rings. (*Note*: The use of an internally guided expansion joint does not eliminate the necessity of using adequate external pipe guides.)

UNIVERSAL EXPANSION JOINT: A universal expansion joint contains two bellows by a common connector for the purpose of absorbing any combination of the three basic movements, that is, axial movement, lateral deflection, and angular rotation. Universal expansion joints are usually furnished with limit rods to distribute the movement between the two bellows of the expansion joint and stabilize the common connector. This definition does not imply that only a double bellows expansion joint can absorb universal movement.

HINGED EXPANSION JOINT: A hinged expansion joint contains one bellows and is designed to permit angular rotation in one plane only by the use of a pair of pins through hinge plates attached to the expansion joint ends. The hinges and hinge pins must be designed to restrain the thrust of the expansion joint due to internal pressure and extraneous forces, where applicable. Hinged expansion joints should be used in sets of two or three to function properly.

SWING EXPANSION JOINT: A swing expansion joint is designed to absorb lateral deflection and/or angular rotation in one plane. Pressure thrust and extraneous forces are restrained by the use of a pair of swing bars, each of which is pinned to the expansion joint ends.

GIMBAL EXPANSION JOINT: A gimbal expansion joint is designed to permit angular rotation in any plane by the use of two pairs of hinges affixed to a common floating gimbal ring. The gimbal ring, hinges, and pins must be designed to restrain the thrust of the expansion joint due to internal pressure and extraneous forces, where applicable.

PRESSURE BALANCED EXPANSION JOINT: A pressure balanced expansion joint is designed to absorb axial movement and/or lateral deflection while restraining the pressure thrust by means of tie devices interconnecting the flow bellows with an opposed bellows also subjected to line pressure. This type of expansion joint is normally used where a change of direction occurs in a run of piping. The flow end of a pressure balanced expansion joint sometimes contains two bellows separated by a common connector, in which case it is called a universal pressure balanced expansion joint.

PRESSURE THRUST FORCE

The static thrust F_s due to internal pressure is given by Eq. 5.4:

$$F_s = ap \qquad (5.4)$$

where a = effective area corresponding to the mean diameter of the corrugations, sq in.
p = design line pressure based on most severe condition, psi
The force required to compress the expansion joint in the axial direction F_m is:

$$F_m = \text{(axial spring constant)(amount of compression)} \qquad (5.5)$$

The centrifugal thrust F_ρ at the elbow due to flow is given by:

$$F_\rho = \frac{2A\rho V^2}{g} \sin \frac{\theta}{2} \qquad (5.6)$$

where A = internal area of pipe, sq in.
ρ = density of fluid, lb/ft^3
V = velocity of flow, ft/sec
g = acceleration due to gravity = 32.2 ft/sec^2
θ = angle of bend
Figure 5.17 shows the elbow where a main anchor is located. The design

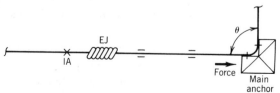

FIGURE 5.17   Anchor force at elbow.

anchor force should include pressure thrust, centrifugal thrust, friction at supports and guides, and force to compress the bellows.

Example

Using the EJMA (reference 1) equation, calculate hydrostatic examination test pressure if the design pressure is 125 psig and design temperature is 500°F. The bellows material is carbon steel ASTM A53 Grade B.

The test pressure is: (using Eq. 2.7)

$$P_t = 1.5 \frac{P_d S_t}{S_d}$$

where P_d = design pressure = 125 psig
S_t = allowable stress of bellows material at test pressure (70°F) = 20,000 psi (S_c from Appendix A3)
S_d = allowable stress of bellows material at design temperature of 500°F
= 18,900 psi (S_h from Appendix A3)

Thus $P_t = 1.5 \left[\dfrac{125(20,000)}{18,900} \right] = 198.4$ psig

EXERCISES

1. (a) Size the expansion loop for the following conditions:
   Diameter = 16 in. standard weight
   Material = A53 Grade A
   Distance between anchors   = 220 ft
   Wt/ft of pipe length   = 80 lb
   Temperature = 750°F   Span = 25 ft
   (b) Calculate the force at anchors for shoes with Teflon slide plate.
   (c) Calculate the force at guides.

2. (a) Design the expansion loop, by equation, with loop height to width ratio as 1.
   Distance between anchors = 225 ft
   Temperature = 800°F Span = 20 ft
   Diameter = 12 in. standard weight
   Material = A53 Grade B
   (b) Calculate the force at anchors for shoes with steel on steel.
   (c) Calculate the force at guides.

3. (a) Calculate the thermal expansion at A and B in the piping system given in Figure 5.18. Material A106 Grade B at 750°F.

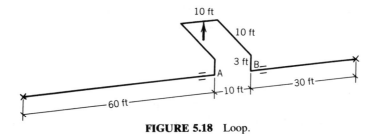

FIGURE 5.18 Loop.

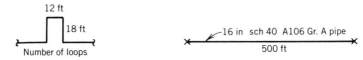

FIGURE 5.19 Number of loops.

(b) Which of the following is advantageous to use: (1) symmetrical expansion loop? (2) unsymmetrical expansion loop?

4. The dimension of an expansion loop is limited as shown in Figure 5.19. If a pipe has a temperature of 650°F, how many expansion loops are required for 500 ft long pipe?

5. If a line is anchored at both ends, but anchors have thermal movement as shown in Figure 5.20, what is the size of the loop? It is 4 in. sch 80, A53 Grade B carbon steel pipe at 350°F.

6. A 6 in. diameter loop has standard sch A53 Grade B pipe with operating temperature 375°F.

   For loop shown in Figure 5.21: (a) find resultant force F at anchors; (b) find moment M at anchors.

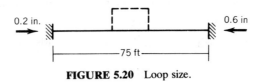

FIGURE 5.20 Loop size.

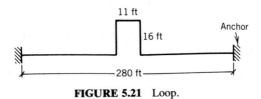

FIGURE 5.21 Loop.

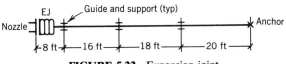

FIGURE 5.22 Expansion joint.

7. Size an expansion loop based on the following conditions: a 12 in. A53 Grade B sch 40 pipe; temperature is 350°F, loop width is 8 ft; and length of pipe is 180 ft

8. From manufacturer's catalog find overall length of flexible hose needed for $\pm\frac{1}{2}$ in. offset deflection for a 6 in. internal diameter hose. Assume type of end connection.

9. A 12 in. diameter carbon steel standard weight pipe is at 525°F. Design pressure is 180 psig. With a single bellows expansion joint in the piping system in Figure 5.22, calculate forces at nozzle and anchor.

   The mean area of convolution is 151 sq in.; the axial spring rate is 882 lb/in.

10. A 40 in. diameter turbine exhaust duct system is fabricated of $\frac{5}{8}$ in. wall carbon steel and operates at full vacuum at 320°F. The movement at the turbine exhaust flange and the condenser inlet are determined as shown in Figure 5.23. A universal pressure-balanced expansion joint is located between two pieces of equipment with the dimensions as shown in Figure 5.23. Determine the forces and moments due to the bellows stiffness at the condenser and turbine connections. The data provided by the expansion joint manufacturer are as follows:
    Mean diameter of bellows $d_p = 42$ in.
    Working spring rate $f_w = 32,000$ lb/in./convolution
    Number of convolutions flow bellows $N_f = 6+6$
    Number of convolutions balancing bellows $N_b = 6$

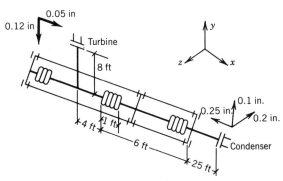

FIGURE 5.23 Universal pressure-balanced expansion joint.

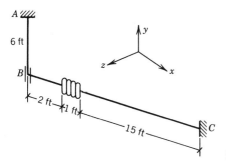

FIGURE 5.24 Single bellows expansion joint.

11. A single bellow expansion joint is placed in a 20 in. diameter carbon steel pipe that runs between anchors A, B, C. Anchor point B is actually a directional guide that restricts only the axial movement. The line is operating at 150 psig and 550°F. Pipe lengths are shown in Figure 5.24. What are the forces and moments acting at A, B, C? The data provided by the expansion joint manufacturer are:

   Effective area corresponding to bellows mean diameter = 480 in.^{2}
   Mean diameter $d_p = 21.5$ in.
   Working spring rate $f_w = 24{,}800$ lb/in./convolution
   Bellows free length = 12 in.
   Number of convolutions $N = 12$

REFERENCES

1. Expansion Joint Manufacturers Association 1973 Addenda to Standards of EJMA, 3rd ed., 1969.
2. Robert L. Benson, Chemetron Corp. "A Basic to Analyzing Piping Flexibility," *Chemical Engineering* (Oct. 23, 1973).
3. Engineering data on expansion joints are available from (company or trade name): Pathway, Flexonics, Adsco, Solar, Anaconda, Temp. Flex, Tube turns, Zallea Bros., and Metal Bellows.

CHAPTER SIX

FLANGED JOINTS

Flanges are used to join sections of pipe lengths and to connect piping to equipments. Two main types of flanges are flat face and raised face. In pipe stress analysis, the capability of a flange to carry external moment is given importance. The actual design of flanged joints can be obtained from other sources (references 1 and 2).

The effects of bolt preload, pressure, temperature, and external moments are discussed below.

Bolt Preload: The initial tightening of the bolt is a prestressing operation. The amount of initial bolt stress developed should be enough to provide against all conditions that tend to produce a leaking joint and at the same time not so excessive that the yielding of the bolts or flanges can produce relaxation that can also result in leakage. For the joint to be tight under hydrostatic (one and half times the design pressure) pressure, an initial bolt stress higher than the design stress value may be allowed.

Internal Pressure: When internal pressure is applied, further yielding of bolt may cause leakage if the margin between initial bolt stress and yield strength is less.

External Pressure: The combined force of external bending moment and bolt loading may plastically deform certain gaskets that result in loss of gasket pressure when the connection is depressurized.

Temperature: Increase in temperature reduces the pressure to which the flange can be subjected. At elevated temperatures, the design stress values are governed by creep rate. If the coefficient of thermal expansion is different (different material) for flange and bolts, leakage may occur due to increase in bolt load. Then retightening of the bolt may be necessary, but it must not be forgotten that the effects of repeated retightening can be cumulative and may make the joint unserviceable.

NOMENCLATURE

S_y yield stress of flange material, psi

C bolt circle diameter of flange, inches

A_b total cross-sectional area of bolts at root of thread, sq in.

D_o outside diameter of flange raised face

PR pressure concurrent with bending moment under dynamic loading

G diameter of location of gasket load reaction, inches (can be approximated by inside diameter of flange raised face)

S_a allowable bolt stress, psi

Units: moments   ft-lb

        stress   psi

OBE   operating basis earthquake

SSE   safety shutdown earthquake

SAM   sesimic anchor movement

Faulted condition is associated with SSE or pipe break. It is an extremely low probability event.

LOCA   Loss of coolant accident. The result would be an inadvertent opening of the pressurized safety or relief valve because of the loss of coolant in excess of the capacity of the reactor coolant make up system.

EXTERNAL MOMENTS

The effect of external moments will be discussed in detail. The allowable moments can be calculated by the three methods outlined by ASME Section III, Nuclear Power Plants Components Code NC-3658.

Method 1: This refers to ANSI B16.5 flanged joints with high strength bolting (bolt material with allowable stress at 100°F not less than 20,000 psi).

    (a)   For service levels A and B under static loads given by Eq. 6.1

    (b)   For service levels A and B under static and dynamic loads in Eq. 6.2

    (c)   For service levels C and D under static and dynamic loads in Eq. 6.3

Method 2: This method concerns standard flanged joints at moderate pressures and temperatures in ANSI B16.5, MSS SP-44, API 605 standards (pressure less than 100 psi and temperature less than 200°F).

Method 3: This is the equivalent pressure method.

Levels A and B service limits must be satisfied for all loadings identified in the design specification in the performance of its specified service function. The component or support must withstand these loadings without damage requiring repair.

Levels C and D service limits permit large deformations in areas of structural discontinuity. The occurrence of stress to level C and D limit may necessitate the removal of the component from service for inspection or repair of damage to the component or support.

COMPARISON OF ALLOWABLE AND ACTUAL MOMENTS

Allowable Moments

Method 1: (high strength bolting option) the design limits and service level limits A and B are:

$$M_{\text{allow static}} = 3125 \left(\frac{S_y}{36,000}\right) CA_b \left(\frac{1}{12}\right) \qquad (6.1)$$

where $S_y/36,000$ should not be greater than unity.

$$M_{\text{allow dynamic}} \text{ (for level } A \text{ or } B) = 6250 \left(\frac{S_y}{36,000}\right) CA_b \left(\frac{1}{12}\right) \qquad (6.2)$$

As can be seen, the results of Eq. 6.2 will be two times that of Eq. 6.1. The design limits and service level limits C and D (faulted) are:

$$M_{\text{allow dynamic}} \text{ (for } C \text{ or } D) = [11{,}250 A_b - \frac{\pi}{16} (D_o^2 PR)] \frac{C}{12} \left(\frac{S_y}{36,000}\right) \qquad (6.3)$$

In method 2: (for flanges at moderate pressures and temperatures)

$$M_{\text{allow static}} = A_b C \left(\frac{S_a}{4}\right) \frac{1}{12} \qquad (6.4)$$

$$M_{\text{allow dynamic}} = A_b C \left(\frac{S_a}{2}\right) \frac{1}{12} \qquad (6.5)$$

In method 3: (equivalent pressure method)

$$P_{\text{eq}} = \frac{16(12)M}{\pi G^3} \qquad (6.6)$$

where M is the largest moment (actual) from Eqs. 6.7, 6.8, and 6.9.

$$\text{Actual or total pressure} = P_{\text{eq}} + \text{design pressure} \qquad (6.6a)$$

To qualify the flange under this method,

P_{eq} plus design pressure should be less than the rated pressure   (6.6b)

Actual Moments

$M(\text{normal}) = M_{\text{actual static}} =$ higher of torsional or resultant of two bending moments for gravity plus thermal normal loading, sustained anchor movement plus relief valve thrust force and other mechanical sustained loads.   (6.7)

$M(\text{upset}) = M_{\text{actual dynamic}} =$ higher of torsional or resultant of two bending moments plus thermal upset plus OBE plus SAM OBE plus LOCA   (6.8)

$M(\text{faulted}) = M_{\text{actual dynamic (faulted)}} =$ higher of torsional or two resultant bending moments plus thermal upset plus SSE plus SAM SSE plus LOCA   (6.9)

$M =$ greater of the above three actual moments   (6.10)
This moment will be used to get equivalent pressure.

As can be expected, for approving the use of the flange at certain locations the actual or calculated bending moments must be lower than the allowable moments. Table 6.1 gives the equation numbers for the actual and the allowable moments for comparison.

Gaskets: Section NC-3647.5 allows only metallic or asbestos gaskets if the expected normal service pressure exceeds 720 psi or the temperature

TABLE 6.1   **Comparison of Allowables Against Actuals**

Actual	Allowables
$M(\text{normal})$ (Eq. 6.7)	$M_{\text{allow static}}$ (Eq. 6.1)
$M(\text{upset})$ (Eq. 6.8)	$M_{\text{allow dynamic}}$ (Eq. 6.2)
$M(\text{faulted})$ (Eq. 6.9)	$M_{\text{allow dynamic}}$(faulted) (Eq. 6.3)
$M(\text{normal})$ (Eq. 6.7)	$M_{\text{allow static}}$ (Eq. 6.4)
$M(\text{upset})$ (Eq. 6.8)	$M_{\text{allow dynamic}}$ (Eq. 6.5)
$M(\text{faulted})$ (Eq. 6.9)	$M_{\text{allow dynamic}}$ (Eq. 6.5)
P_{eq} + design pressure (Eq. 6.6a)	Rated pressure

exceeds 750°F. However, compressed sheet asbestos-confined gaskets are not limited as to pressure provided the gasket material is suitable for the temperature.

Example Problem

Calculate the allowable and actual bending moments and check if the given flange is qualified according to ASME Section III, NC-3658 (summer 1979).

Pipe diameter = 30 in.

The OD of the flange raised face = 33.75 in.

Number of bolts = 28

Total bolt area = 28(0.8898) = 24.94 sq in.

Diameter of bolt circle C = 36 in.

The flange material is carbon steel SA105
The bolt material is SA193 Grade B 7
Bolt allowable stress = 25,000 psi
Flange material yield stress S_y = 32,800 psi
Pressure rating = 150 psi
Design temperature = 200°F
Design pressure = 175 psi

Actual moments (ft-lb) from piping analysis is given in Table 6.2.
The higher of the torsional moment or resultant bending moment is

TABLE 6.2   Actual Moments from Piping Analysis (ft-lb)

Loading	M_T	BM_1	BM_2	$M_R = \sqrt{BM_1^2 + BM_2^2}$	Higher of M_T or M_R
Dead weight	1,084	1,939	11,520	11,682	11,682
Thermal	1,901	6,350	2,825	6,950	6,950
OBE	8,518	7,979	9,817	12,650	12,650
OBE SAM	0	0	0	0	0
SSE	18,354	16,638	10,448	19,646	19,646
SSE SAM	0	0	0	0	0
LOCA	0	0	0	0	0

$M_{\text{actual static}}(\text{normal}) = 11,682 + 6950 = 18,632$ ft-lb (from Eq. 6.7)
$M_{\text{actual dynamic}}(\text{upset}) = 11,682 + 6950 + 12,650 = 31,282$ (from Eq. 6.8)
$M_{\text{actual dynamic}}(\text{faulted}) = 11,682 + 6950 + 19,646 = 38,278$ (from Eq. 6.9)

tabulated in Table 6.2. Equations 6.7, 6.8, and 6.9 are used to calculate total actual moment for normal, upset, and faulted conditions.

ALLOWABLE MOMENTS

The bolt material is SA193 Grade B7 alloy steel with allowable stress 25,000 psi.

Method 1, known as high strength bolting option, is used because the bolt allowable stress is greater than 20,000 psi at 100°F. Thus Eqs. 6.1, 6.2, and 6.3 are used to calculate allowable moments.

$$M_{\text{allow static}} = 3125 \left(\frac{32,800}{36,000}\right)\left(\frac{36}{12}\right)(24.92) = 212,858 \text{ ft-lb} \quad (6.1)$$

$$M_{\text{allow dynamic}} = 2(212,858) = 425,716 \text{ ft-lb} \quad (6.2)$$

$$M_{\text{allow dynamic}}(\text{faulted}) = \left[(11,250)(24.92) - \frac{\pi}{16}(33.75)^2(175)\right]\left(\frac{36}{12}\right)\frac{32.8}{36}$$

$$= 659,310 \text{ ft-lb} \quad (6.3)$$

Table 6.3 gives the comparison of moments of the Example Problem.

The effect of flange material, flange rating, and flange diameter on allowable moment is shown in Table 6.4. Internal pressure at flange is 175 psi. As can be expected the allowable moments are higher for larger flanges and higher ratings. The allowable moments for carbon steel flanges are higher than for stainless steel flanges because yield stress (used in high strength bolting option) for carbon steel is higher. The yield strength for carbon steel is 32,800 psi as compared with 21,300 psi for stainless steel at 200°F.

TABLE 6.3   Comparison of Moments (ft-lb)

Condition	Actual Moment	Allowable	Moment
Normal	18,632	212,858	OK
Upset	31,282	425,716	OK
Faulted	38,278	659,310	OK

TABLE 6.4 Flange Allowable Moments (ft-lb) Using Method 1

Pipe Diameter (in.)	Total Bolts Area A_b (sq. in.)	Bolt Circle Diameter (in.)	Rating (psi)	Allowable Moment Static (Eq. 6.1)	Static + Dynamic (Eq. 6.2)	Static + Dynamic (Eq. 6.3)	Flange Material	Bolt Material
2	0.81	4.75	150	912	1,825	3,123	Carbon steel SA105	SA193 Grade B7
2	0.81	4.75	150	592	1,185	2,028	Stainless steel SA 182	SA193 Grade B7
3	0.81	6	150	748	1,497	2,441	Grade F304	SA 193 Grade B7
4	1.62	7.5	150	1,872	3,744	6,252	Grade F304	SA193 Grade B7
4	2.42	7.88	300	4,524	9,049	15,501	Grade F304	SA193 Grade B7
8	2.42	11.75	150	6,746	13,493	20,825	Carbon steel SA105	SA193 Grade B7
8	5.03	13.0	300	15,515	31,030	52,022	Stainless steel	SA 193 Grade B7

The temperature is 200°F and the pressure is 175 psi.
S_y for carbon steel at 200°F is 32,800 psi. S_y for stainless steel at 200°F is 21,300 psi.

REFERENCES

1. ASME Sec. III, Div. 1 code. "Nuclear Power Plant Components," Article XI-3000.
2. ASME Sec. VIII, Div. 1 code. "Design of Flanged Joints," Appendix II.
3. ASME Sec. III, Div. 1 code. "Nuclear Power Plant Components," subsection NC-3658 (summer 1979).
4. "Flange Qualification Program," *Tennessee Valley Authority*.
5. ANSI B16.5. "Steel Pipe Flanges and Flanged Fittings" (1977).
6. API 605. Reaffirmed in 1973, "Large Diameter Carbon Steel Flanges."

CHAPTER SEVEN

PIPING CONNECTED TO NONROTATING EQUIPMENT

The external loads imposed on nonrotating equipment by piping should be below the allowable loads supplied by equipment manufacturers. Examples of nonfired equipments are heat exchangers, tanks, pressure vessels, drums, air coolers, and condensers. Examples of fired vessels are boilers and fires heaters.

The actual forces and moments from piping stress analysis may be sent to manufacturers to get these loads approved.

The methods to calculate local stresses on the vessel and nozzle intersection are:

1. Finite element analysis that is more accurate but could be expensive for computer resources.
2. Local stress calculation outlined by Welding Research Council (WRC) bulletin 107 (reference 1).
3. Local stress calculations using Flügge–Conrad solutions (reference 2).
4. WRC bulletin 297, (reference 8) Local Stress in Cylindrical Shells, supplement to WRC bulletin 107.

For each piece of equipment, applicable code and standard requirements should be satisfied. Instead of reprinting text information available from other sources, a discussion with specific examples for cylindrical and spherical vessels is presented here.

LOCAL STRESS CALCULATION USING WRC 107 BULLETIN

Based on work done by Bijlaard, WRC 107 was prepared. Sign conventions used are exactly as given in the bulletin.

Comments on Using WRC 107 Bulletin

1. Vessel (cylindrical) diameter to vessel thickness ratio range is $10 \leq D/T \leq 600$.
2. Nozzle diameter to vessel diameter ratio range is $0.02 \leq d/D \leq 0.57$.
3. Nozzle thickness is not considered for cylindrical vessel.
4. Nondimensional constants read from curve from WRC 107 bulletin are for acceptable ranges only. Extensions of curves can be used only if allowed. Values outside the range may give unconservative results.
5. March 1979 revision of the bulletin gives important revisions. Earlier versions should be carefully used.
6. Signs for stress were obtained by considering the deflection of shell resulting from the various modes of loadings. Tensile stress is marked as $+$ and compressive stress is marked as $-$.
7. Maximum shear theory has been used to determine stress intensity.
8. Welding Research Council 107 omits the internal pressure stress. The effect of pressure may be included if desired.
9. The stresses calculated are in the vessel wall (shell) and not in the nozzle. Stresses may be higher in the nozzle wall in case the nozzle opening is not reinforced.
10. Welding Research Council 107 method may be used for ellipsoidal heads as well as cylindrical and spherical shells.
11. Stresses due to radial load in cylindrical shells are not applicable if the length of the cylinder is less than its radius. The curves are for length radius ratio of 8.
12. Stresses due to external moment are not applicable if the attachment is located within the distance of half the shell radius from the near end of the wall.

Table 7.1 gives stress concentration factors K_n and K_b. The equations for calculating the stress concentration factors K_n and K_b are given in Eqs. 7.1 and 7.2. Table 7.1 was generated using Eqs. 7.1 and 7.2.

The actual stress calculated is compared with the allowable stress. If the actual stress is higher, a pad thickness is assumed and the calculation is rerun with the total thickness (sum of vessel and pad thickness) as vessel thickness. In practice, the assumed pad thickness is equal to the vessel thickness. If double the thickness is not enough, efforts may be made to reduce the loadings on the vessel.

$$\text{For membrane load,} \qquad K_n = 1 + \left[\frac{1}{5.6\dfrac{r}{2T}} \right]^{0.65} \qquad (7.1)$$

$$\text{For bending,} \qquad K_b = 1 + \left[\frac{1}{9.4\dfrac{r}{2T}}\right]^{0.8} \qquad (7.2)$$

where r = radius used for nozzle-to-shell interface (in.) and T = shell thick-ness (in.).

TABLE 7.1   Concentration Factors
(Based on $\frac{3}{8}$-in. radius at shell-to-nozzle interface)

T (in.)	K_n	K_b
$\frac{3}{8}$	1.5121	1.2899
$\frac{7}{16}$	1.5661	1.3280
$\frac{1}{2}$	1.6174	1.3650
$\frac{9}{16}$	1.6665	1.4010
$\frac{5}{8}$	1.7138	1.4363
$\frac{11}{16}$	1.7594	1.4709
$\frac{3}{4}$	1.8036	1.5048
$\frac{7}{8}$	1.8882	1.5711
1	1.9688	1.6355
$1\frac{1}{8}$	2.0459	1.6983
$1\frac{1}{4}$	2.1200	1.7597
$1\frac{1}{2}$	2.2609	1.8790
$1\frac{3}{4}$	2.3938	1.9943
2	2.5202	2.1064

Example Problems

1. Calculate the local stress for the *cylindrical vessel* given as follows (reference 1):
    Vessel radius = R_m = 72 in.
    Vessel thickness T = 0.4375 in.
    Attachment radius r_o = 3.125 in.
    Geometric parameters are:

    $$v = \frac{R_m}{T} = \frac{72}{0.4375} = 164$$

    $$\beta = 0.875\frac{r_o}{R_m} = 0.875\frac{3.125}{72} = 0.043$$

The stress concentration factors are for the membrane load $K_n = 1.0$, for the bending load $K_b = 1.0$.

The applied loads are:

Radial load $P = -97.8$ lb

Circular moment $M_c = -768$ in.-lb

Longitudinal moment $M_1 = -10,152$

Torsional moment $M_T = 31,368$

Shear load $V_c = -4$

Shear load $V_1 = -451$

The nondimensional constants read from graphs of WRC 107 are:

WRC 107 Graph Number	Constant	Value
3C	$\dfrac{N_\phi}{R/R_m}$	30.0
1C	$\dfrac{M_\phi}{P}$	0.15
3A	$\dfrac{N_\phi}{M_c/R_m^2\beta}$	4.1
1A	$\dfrac{M_\phi}{M_c/R_m\beta}$	0.095
3B	$\dfrac{N_\phi}{M_L/R_m^2\beta}$	16.0
1B or 1B-1	$\dfrac{M_\phi}{M_L/R_m\beta}$	0.048
4C	$\dfrac{N_x}{P/R_m}$	30.0
2C	$\dfrac{M_x}{P}$	0.095
4A	$\dfrac{N_x}{M_c/R_m^2\beta}$	5.6
2A	$\dfrac{M_x}{M_c/R_m^2\beta}$	0.055
4B	$\dfrac{N_x}{M_L/R_m^2\beta}$	4.2
2B or 1B-1	$\dfrac{M_x}{M_L/R_m\beta}$	0.075

2. Calculate local stress for the spherical vessel given as follows:

Vessel mean radius $= 167.43$ in.
Vessel thickness $= 1.125$
Nozzle thickness $= 0.5$
Nozzle mean radius $= 11.75$
Nozzle outside radius $= 12$
The applied loads are:
Radial load $= p = 1977$ lb
Shear load $= V_1 = 97$
Shear load $= V_2 = -36$
Overturning moment $= M_1 = -158,808$ in.-lb
Overturning moment $= M_2 = -47,976$
Torsional moment $= M_T = -10,344$
The concentration factors are: $K_n = 2.0$ and $K_b = 2.0$.
The geometric parameters are:

$$\nu = \frac{r_m}{t} = \frac{11.75}{0.5} = 23.5$$

$$\rho = \frac{T}{t} = \frac{1.125}{0.5} = 2.25$$

$$U = \frac{r_o}{\sqrt{R_m T}} = \frac{12}{\sqrt{167.43 \times 1.125}} = 0.8746$$

Nondimensional constants from WRC 107 graphs are as follows:

SP 1 to SP 10		SM 1 to SM 10			
$\dfrac{N_x T}{p}$	$\dfrac{M_x}{p}$	$\dfrac{N_x T \sqrt{R_m T}}{M_1}$	$\dfrac{M_x \sqrt{R_m T}}{M_1}$	$\dfrac{N_x T \sqrt{R_m T}}{M_2}$	$\dfrac{M_x \sqrt{R_m T}}{M_2}$
0.0078	0.04	0.03	0.09	0.083	0.25

SP 1 to SP 10		SM 1 to SM 10			
$\dfrac{N_y T}{p}$	$\dfrac{M_y}{p}$	$\dfrac{N_y T \sqrt{R_m T}}{M_1}$	$\dfrac{M_y \sqrt{R_m T}}{M_1}$	$\dfrac{N_y T \sqrt{R_m T}}{M_2}$	$\dfrac{M_y \sqrt{R_m T}}{M_2}$
0.055	0.025	0.119	0.068	0.33	0.189

In an effort to extend WRC 107 results to larger D/T and smaller d/D valves and to include the effect of the nozzle thickness, calculation using Flügge–Conrad solutions is presented (reference 2).

WRC Bulletin 297 (reference 8) broadens the coverage of WRC Bulletin 107 and is based on Steele's theory (reference 2). WRC 297 includes the

TABLE 7.2 Computation Sheet for Local Stresses in Cylindrical Shells (Reference 1)

1. Applied Loads*
Radial load	$P = -97.8$ lb
Circ. moment	$M_c = -768$ in. lb
Long. moments	$ML = -10{,}152$ in. lb
Torsion moment	$M_t = 31{,}368$ in. lb
Shear load	$V_c = -4$ lb
Shear load	$VL = -451$ lb

2. Geometry
Vessel thickness	$T = 0.4375$ in.
Attachment radius	$r_o = 3.125$ in.
Vessel radius	$R_m = 72$ in.

3. Geometric Parameters

$$\gamma = \frac{R_m}{T} = \frac{72}{0.4375} = 164$$

$$\beta = (0.875)\,\frac{r_o}{R_m} = 0.043$$

4. Stress Concentration due to:
   a) membrane load, $K_n = 1.0$
   b) bending load, $K_b = 1.0$
   *Note: Enter all force values in accordance with sign convention

(Diagram labels: P, M_L, M_c, M_r, V_L, θ, v_c, r_e, $\frac{L}{2}$, B_U, B_L, C_U, C_L, D_U, D_L, A_U, A_L, T, R_m, Round attachment, CYLINDRICAL SHELL)

From Fig.	Read Curves for	Compute Absolute Values of Stress	STRESSES psi. If Load Is Opposite That Shown, Reverse Signs Shown							
			A_u	A_L	B_u	B_L	C_u	C_L	D_u	D_L
3C or 4C	$\dfrac{N_\phi}{P/R_m}$	$K_n\left(\dfrac{N_\phi}{P/R_m}\right)\dfrac{P}{R_m T}$	-93	-93	-93	-93	-93	-93	-93	-93
1C or 2C-1	$\dfrac{M_\phi}{P}$	$K_b\left(\dfrac{M_\phi}{P}\right)\dfrac{6P}{T^2}$	-460	$+460$	-460	$+460$	-460	$+460$	-460	$+460$
3A	$\dfrac{M_c}{R_m^2\beta}$	$K_n\left(\dfrac{N_\phi}{M_c/R_m^2\beta}\right)\dfrac{M_c}{R_m^2\beta T}$					-32	-32	$+32$	$+32$
1A	$\dfrac{M_\phi}{M_c/R_m\beta}$	$K_b\left(\dfrac{M_\phi}{M_c/R_m\beta}\right)\dfrac{6M_c}{R_m\beta T^2}$					-740	$+740$	$+740$	-740
3B	$\dfrac{N_\phi}{ML/R_m^2\beta}$	$K_n\left(\dfrac{M_\phi}{ML/R_m^2\beta}\right)\dfrac{ML}{R_m^2\beta T}$	-1667	-1667	$+1667$	$+1667$				
1B or 1B-1	$\dfrac{M_\phi}{ML/R_m\beta}$	$K_b\left(\dfrac{M_\phi}{ML/R_m\beta}\right)\dfrac{6ML}{R_m\beta T^2}$	-4945	$+4945$	$+4945$	-4945				

Add algebraically for summation of φ stresses, σ_φ =

Add algebraically for summation of φ stresses, $\sigma_\phi =$		-7165	3644	6058.5	-3750	-905	1074	218	-341
3C or 4C	$\dfrac{N_x}{P/R_m}$ $\quad$ $K_n\left(\dfrac{N_x}{P/R_m}\right)\dfrac{P}{R_mT}$	-93	-93	-93	-93	-93	-93	-93	-93
1C-1 or 2C	$\dfrac{M_x}{P}$ $\quad$ $K_b\left(\dfrac{M_x}{P}\right)\dfrac{6P}{T^2}$	-291	$+291$	-291	$+291$	-291	$+291$	-291	$+291$
4A	$\dfrac{N_x}{M_c/R_m^2\beta}$ $\quad$ $K_n\left(\dfrac{N_x}{M_c/R_m^2\beta}\right)\dfrac{M_c}{R_m^2\beta T}$					-44	-44	$+44$	$+44$
2A	$\dfrac{M_x}{M_c/R_m\beta}$ $\quad$ $K_b\left(\dfrac{M_x}{M_c/R_m\beta}\right)\dfrac{6M_c}{R_m\beta T^2}$					-428	$+428$	$+428$	-428
4B	$\dfrac{N_x}{ML/R_m^2\beta}$ $\quad$ $K_n\left(\dfrac{N_x}{ML/R_m^2\beta}\right)\dfrac{ML}{R_m^2\beta T}$	-437	-437	$+437$	$+437$				
2B or 2B-1	$\dfrac{M_x}{ML/R_m\beta}$ $\quad$ $K_b\left(\dfrac{M_x}{ML/R_m\beta}\right)\dfrac{6ML}{R_m\beta T^2}$	-7726	$+7726$	$+7726$	-7726				
Add algebraically for summation of X stresses, $\sigma_x =$		-8547	$+7487$	7779	-7673	-856	$+582$	$+88$	-186
Shear stress due to torsion, M_T	$\tau\phi = \tau x\phi = \dfrac{M_T}{2\pi r_o^2 T}$	$+1173$	$+1173$	$+1173$	$+1173$	$+1173$	$+1173$	$+1173$	$+1173$
Shear stress due to load, V_c	$\tau x\phi = \dfrac{V_c}{\pi r_o T}$	$+0.9$	$+0.9$	-0.9	-0.9				
Shear stress due to load, V_L	$\tau x\phi = \dfrac{V_L}{\pi r_o T}$					-105	-105	$+105$	$+105$
Add algebraically for summation $\tau =$		1174	1174	1172.7	1172	1068	1068	1278	1278

Combined stress intensity, S

1. When $\tau \neq 0$, $S =$ largest absolute magnitude of either

$$= \tfrac{1}{2}\left[\sigma_x + \sigma_\phi \pm \sqrt{(\sigma_x - \sigma_\phi)^2 + 4\tau^2}\right] \quad \text{or} \quad \sqrt{(\sigma_x - \sigma_\phi)^2 + 4\tau^2}$$

$$= \tfrac{1}{2}\left[-8547 - 7165 \pm \sqrt{(-8547 + 7165)^2 + 4(1174)^2}\right] = 9218 \text{ psi.}$$

2. When $\tau = 0$, $S =$ largest absolute magnitude of either

$$= \sigma_x,\ \sigma_\phi \text{ or } (\sigma_x - \sigma_\phi).$$

From WRC 107, ASME.

TABLE 7.3 Computation Sheet for Local Stresses in Spherical Shells (Hollow Attachment) (Reference 1)

1. Applied Loads*

Radial load	$P = 1977\ \text{lb}$
Shear load	$V_1 = 97\ \text{lb}$
Shear load	$V_2 = -36\ \text{lb}$
Overturning moment	$M_1 = 158{,}808\ \text{in. lb}$
Overturning moment	$M_2 = -49{,}976\ \text{in. lb}$
Torsional moment	$M_T = -10{,}344\ \text{in. lb}$

2. Geometry

Vessel thickness	$T = 1.125\ \text{in.}$
Vessel mean radius	$R_m = 167.43\ \text{in.}$
Nozzle thickness	$t = 0.5\ \text{in.}$
Nozzle mean radius	$r_m = 11.75\ \text{in.}$
Nozzle outside radius	$r_o = 12\ \text{in.}$

3. Geometric Parameters

$$\gamma = \frac{r_m}{t} = \frac{11.75}{0.3} = 23.5$$

$$\rho = \frac{T}{t} = \frac{1.125}{0.5}$$

$$U = \frac{r_o}{\sqrt{R_m T}} = \frac{12}{\sqrt{167.43 \times 1.125}} = 0.875$$

4. Stress Concentration Factors due to:
   membrane load, $K_n = 2.0$
   bending load, $K_b = 2.0$

Note: Enter all force values in accordance with sign convention

HOLLOW ATTACHMENT

From Fig.	Read Curves for	Compute Absolute Values of Stress	STRESSES psi, If Load Is Opposite That Shown, Reverse Signs Shown							
			A_u	A_L	B_u	B_L	C_u	C_L	D_u	D_L
SP 1 to 10	$\dfrac{N_x T}{P}$	$K_n\left(\dfrac{N_x T}{P}\right)\dfrac{P}{T^2}$	−24.4	−24.4	−24.4	−24.4	−24.4	−24.4	−24.4	−24.4
	$\dfrac{M_x}{P}$	$K_b\left(\dfrac{M_x}{P}\right)\dfrac{6P}{T^2}$	−637.3	+637.3	−637.3	+637.3	−637.3	+637.3	−637.3	+637.3
SM 1 to 10	$\dfrac{N_x T\sqrt{R_m T}}{M_1}$	$K_n\left(\dfrac{N_x T\sqrt{R_m T}}{M_1}\right)\dfrac{M_1}{T^2\sqrt{R_m T}}$					−548.6	−548.6	+548.6	+548.6
	$\dfrac{M_x\sqrt{R_m T}}{M_1}$	$K_b\left(\dfrac{M_x\sqrt{R_m T}}{M_1}\right)\dfrac{6M_1}{T^2\sqrt{R_m T}}$					−9874	+9874	+9874	−9874
	$\dfrac{N_x T\sqrt{R_m T}}{M_1}$	$K_n\left(\dfrac{N_x T\sqrt{R_m T}}{M_2}\right)\dfrac{M_2}{T^2\sqrt{R_m T}}$	−458.5	−458.5	+458.5	+458.5				

$\dfrac{M_x\sqrt{R_mT}}{M_1}$ $\quad K_b\!\left(\dfrac{M_x\sqrt{R_mT}}{M_2}\right)\dfrac{6M}{T^2\sqrt{R_mT}}$	−8286	−8286	+8286	−8286	−11,084	9931.4	9760	−8713
Add algebraically for summation of σ_x	−9406	8440	8083	−7714	−11,084	9931.4	9760	−8713
SP 1 to 10 $\quad K_n\!\left(\dfrac{N_yT}{P}\right)\dfrac{P}{T^2}$	−171.8	−171.8	−171.8	−171.8	−171.8	−171.8	−171.8	−171.8
$\dfrac{N_xT}{P}$								
$\dfrac{M_y}{P}$ $\quad K_b\!\left(\dfrac{M_y}{P}\right)\dfrac{6P}{T^2}$	−468.6	−468.6	+468.6	+468.6	−468.6	+468.6	−468.6	−468.6
SM 1 to 10 $\quad K_n\!\left(\dfrac{N_yT\sqrt{R_mT}}{M_1}\right)\dfrac{M_1}{T^2\sqrt{R_mT}}$								
$\dfrac{N_yT\sqrt{R_mT}}{M_1}$					−2176	−2176	+2176	+2176
$K_b\!\left(\dfrac{M_y\sqrt{R_mT}}{M_1}\right)\dfrac{6M}{T^2\sqrt{R_mT}}$					−7460.4	−7460.4	+7460.4	−7460.4
$\dfrac{M_y\sqrt{R_mT}}{M_1}$								
$K_n\!\left(\dfrac{N_yT\sqrt{R_mT}}{M_2}\right)\dfrac{M_2}{T^2\sqrt{R_mT}}$	−1822.9	−1822.9	+1822.9	+1822.9				
$\dfrac{N_yT\sqrt{R_mT}}{M_2}$								
$K_b\!\left(\dfrac{M_y\sqrt{R_mT}}{M_2}\right)\dfrac{6M_2}{T^2\sqrt{R_mT}}$	−6264	+6264	+6264	−6264				
$\dfrac{M_y\sqrt{R_mT}}{M_2}$								
Add algebraically for summation of σ_y	−8723	−4738	7446	1493	−10,276	5581	8996	−4987
Shear stress due to load V_1 $\quad \tau_1=\dfrac{V_1}{\pi r_oT}$	+0.8	+0.8	−0.8	−0.8	−2.3	−2.3	+2.3	+2.3
Shear stress due to load V_2 $\quad \tau_2=\dfrac{V_1}{\pi r_oT}$								
Shear stress due to torsion M_T $\quad \tau_1=\tau_2=\dfrac{M_T}{2\pi r_o^2T}$	+10.2	+10.2	+10.2	+10.2	+10.2	+10.2	+10.2	+10.2
Add algebraically for summation τ	11	11	9.4	9.4	7.9	7.9	12.5	12.5

Combined stress intensity, S

1. When $\tau \neq 0$, S = largest absolute magnitude of either

$$= \tfrac{1}{2}\!\left[\sigma_x+\sigma_y \pm \sqrt{(\sigma_x-\sigma_y)^2+4\tau^2}\right] \text{ or } \sqrt{(\sigma_x-\sigma_y)^2+4\tau^2}$$

$$= \tfrac{1}{2}\!\left[-11084-10276 \pm \sqrt{(-11084+10276)^2+4(12.5)^2}\right] = 11084 \text{ psi.}$$

2. When $\tau = 0$, S = largest absolute magnitude of either $S = \sigma_x,\ \sigma_y$ or $(\sigma_x-\sigma_y)$.

effect of nozzle thickness and data on nozzle flexibility. Nozzle and vessel are treated as thin-walled cylindrical shells.

ROTATIONAL SPRING RATE FOR CYLINDRICAL VESSEL

As a conservative approach, vessel nozzles are considered rigid in pipe stress calculations. However, the vessel or drum has inherent flexibility that can be advantageously used to obtain lower and more realistic external moments.

Out of three primary forces and three primary moments that may be applied to the shell at a nozzle, only radial force and two moments are considered significant in causing shell deflection. Rotations of elastic ends are usually more significant than translations. Therefore elastic translations of nozzles and thus radial flexibility is ignored.

Inplane and outplane spring rates are important. In a cylindrical vessel, the inplane spring rate corresponds to the longitudinal moment, and the outplane spring rate corresponds to the circumferential moment. As shown in Figure 7.1, by application of the longitudinal bending moment, the plane formed by vessel and nozzle centerlines remains inplane. The circumferential moment will be the outplane moment because this moment will bring nozzle axis into or out of the original plane.

Piping Loads

The point of view that using rotational spring constants result in unconservative values for bending moments should be looked into. It is known that the primary piping loads (weight, pressure) and its effects remain the same in magnitude, whereas secondary loading (thermal) and its effects release itself when the resistance reduces. An example is that, under thermal loading, the bending moments acting on the nozzle drops when the rotation is allowed. All structural systems have inherent flexibility that is expressed as rotational spring rates for vessel and nozzle connections. These spring rates should not be used for pump, turbine, or compressor nozzles.

Equation 7.3 can be used to calculate rotational stiffness at the nozzle connection. Equation 7.4 expresses the formula for calculating flexibility

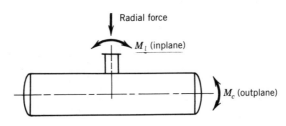

FIGURE 7.1 Vessel and nozzle arrangement.

factor K. Values for constant C is 0.09 for the inplane and C is 0.27 for the outplane bending (reference 3).

$$\frac{M}{\theta} = \frac{EI}{DN\,K}\frac{\pi}{180} \qquad (7.3)$$

where $\dfrac{M}{\theta}$ = spring constant in.-lb/degree

M = bending moment, in.-lb
θ = angle of rotation, radians
E = modulus of elasticity in cold condition, psi
I = area moment of inertia of nozzle, in.^{4}
DN = diameter of nozzle, inches
K = flexibility factor

The flexibility factor K:

$$K = C\left(\frac{D}{T}\right)^{\frac{3}{2}}\left(\frac{TN}{T}\right)\left(\frac{DN}{D}\right) \qquad (7.4)$$

where $C = 0.09$ constant for inplane bending
$C = 0.27$ for outplane bending
D = diameter of vessel, inches
T = wall thickness of vessel, inches
DN = diameter of nozzle, inches
TN = wall thickness of nozzle, inches

These spring rates should not be used if nozzle or pad diameter is greater than one third of the vessel or header diameter.

Example Problem

The following four cases are considered for calculating the spring rate:

1. Vessel nozzle (Fig. 7.2) treated as rigid
2. Using rotational stiffness for a 48 in. diameter vessel

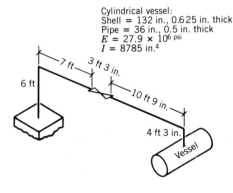

Cylindrical vessel:
Shell = 132 in., 0.625 in. thick
Pipe = 36 in., 0.5 in. thick
$E = 27.9 \times 10^6$ psi
$I = 8785$ in.^{4}

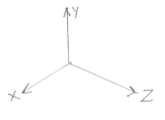

FIGURE 7.2 Piping arrangement for spring rate comparison.

3. Using rotational stiffness for a 60 in. diameter vessel
4. Using rotational stiffness for a 96 in. diameter vessel

The spring constant calculation for the 48 in. vessel, 1 in. thick with 12.75 in. OD nozzle of thickness 0.375 in. follows:

The flexibility factor is:

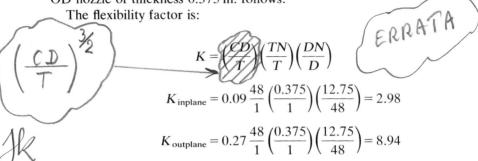

$$K = \left(\frac{CD}{T}\right)\left(\frac{TN}{T}\right)\left(\frac{DN}{D}\right)$$

$$K_{\text{inplane}} = 0.09 \frac{48}{1} \left(\frac{0.375}{1}\right)\left(\frac{12.75}{48}\right) = 2.98$$

$$K_{\text{outplane}} = 0.27 \frac{48}{1} \left(\frac{0.375}{1}\right)\left(\frac{12.75}{48}\right) = 8.94$$

The spring constant is:

$$\frac{M}{\theta} = \frac{EI}{DN} \frac{1}{K} \left(\frac{\pi}{180}\right)$$

$$= \frac{30 \times 10^6 \times 279.3}{12.75(K)} \left(\frac{\pi}{180}\right)$$

The inplane spring constant with K_i as $2.98 = 358 \times 10^4$ in.-lb/deg. The outplane spring constant with K_o as $8.94 = 119 \times 10^4$ in.-lb/deg.

Table 7.4 gives calculated values of spring constants for the cases considered.

The piping stress analysis that was done for the four cases (three different vessel diameters) and the forces and moments at the nozzle are given in Table 7.5. As can be easily seen, the bending moment values in x- and y-axes have dropped when the flexibility of the vessel nozzle is included, say, for the 48 in. diameter as compared to when the flexibility is not included.

TABLE 7.4   Calculated Values of Spring Constants

Vessel Diameter (in.)	Longitudinal Spring Rate (inplane) (in.-lb/deg.)	Circumferential Spring Rate (outplane) (in.-lb/deg.)
48	358×10^4	119×10^4
60	320×10^4	107×10^4
96	253×10^4	844×10^3

TABLE 7.5 Forces and Moments at Vessel Nozzle

	Force (lb)			Moment (ft-lb)		
Case	F_x	F_y	F_z	M_x	M_y	M_z
No flexibility	−481	5,066	−388	25,497	2,248	−583
48 in. dia.	−476	1,866	−466	9,994	2,004	−485
60 in. dia.	−475	1,727	−469	9,321	1,980	−476
96 in. dia.	−474	1,450	−476	7,980	1,918	−451

The example Problem given as Figure 7.2 was selected to compare the values of spring rates calculated here with already published results using a slightly different approach.

In the cylindrical vessel:

Shell 132 in., 0.625 in. thick
Pipe 36 in., 0.5 in. thick
$E = 27.9 \times 10^6$ psi
$I = 8785$ in.^{4}

The outplane flexibility factor is:

$$K = (0.27)\frac{132}{0.625}\frac{0.5}{0.625}\frac{36}{132} = 180.8$$

The outplane spring rate is:

$$\frac{M}{\theta} = \frac{27.9 \times 10^6 \times 8785}{36(180.8)}\frac{\pi}{180} = 776{,}569 \text{ in.-lb/deg.}$$

Table 7.6 gives calculated values of spring rates. The values given are for 132 in. × 0.625 in. thick vessel with 36 in. × 0.5 in. thick pipe. As can be seen the values are close, knowing the values used for rigidity for a system 10^{12} in.-lb/degree.

TABLE 7.6 Comparison of Calculated Spring Rates

Source	Spring Constant (in.-lb/deg.) (outplane)
Stevens, P. G. (reference 5)	727,200
Simplex (reference 6)	659,712
Bijlaard, P. P. (reference 7)	588,252
Eq. 7.3	776,569

The spring rate values when used give reduced bending moments, thus avoiding more piping, nozzle pads, or consideration of alternate arrangements of piping.

Stress range in the vessel shell, which comes under pressure vessel code, is of importance but not discussed here. The equations given here is of assistance when the bending moments are slightly higher than allowed and these slightly higher values can be reduced by using spring rate constants in the analysis.

REFERENCES

1. Wichman, K. R. "Local Stresses in Spherical and Cylindrical Shells due to External Loadings," Welding Research Council Bulletin 107 (revised March 1979).

2. Steel, C. R. "Stress Analysis of Nozzles in Cylindrical Vessels with External Load," *Journal of Pressure Vessel Technology*, Vol. 105 (August 1983).

3. Kannappan S. "Effect of Inclusion of Rotational Spring Rate of Vessel Nozzles in Pipe Stress Calculations," Society of Piping Engineers Conference, Houston in October 1982.

4. Oakridge National Laboratory (ORNL). Phase Report 115-3.

5. Stevens, P. G. et al. "Vessel Nozzle and Piping Flexibility Analysis," *Journal of Engineering for Industry*, May 1962, p. 225.

6. *Simplex Computer Program User's Manual, Peng Engineering*.

7. Bijlaard P. P. "Stresses from Radial Loads and External Moments in Cylindrical Vessel, *The Welding Journal*, Vol. 34 (1955).

8. Mershon, J. L. WRC Bulletin 297, August 1984, Supplement to WRC Bulletin 107.

CHAPTER EIGHT

PIPING CONNECTED TO ROTATING EQUIPMENT

External loads imposed by piping on the rotating equipment nozzles should be less than allowable loads. Examples of rotating equipment are centrifugal pumps, steam turbines, and centrifugal compressors.

If excessive loads are imposed, misalignment may result that affects mechanical operation and could cause objectionable vibration. A close alignment between rotating and stationary parts must be maintained. The provision for expansion of the casing and maintaining close clearances requires that the forces and moments due to the piping are limited.

Instead of duplicating what is available in other sources, examples are given here with references to different standards.

PIPING CONNECTED TO STEAM TURBINES

The NEMA standards SM 23 (reference 1) outlines guidelines for calculating allowable loads.

This standard has two parts:

1. Local allowables at each nozzle
2. Combined allowables for comparing loads transferred to centerline of the exhaust nozzle

The method to transfer forces and moments is given in Eq. 8.1. The following equation is given for two nozzles, but the same equation can be extended further (reference 2):

$$\sum F_x = F_x(\text{inlet}) + F_x(\text{exhaust})$$

$$\sum F_y = F_y(\text{inlet}) + F_y(\text{exhaust})$$

$$\sum F_z = F_z(\text{inlet}) + F_z(\text{exhaust})$$

$$\sum M_x = M_x(\text{inlet}) + M_x(\text{exhaust}) - F_y(\text{inlet})(Z_1) + F_z(\text{inlet})(Y_1) \tag{8.1}$$

$$\sum M_y = M_y(\text{inlet}) + M_y(\text{exhaust}) + F_y(\text{inlet})(Z_1) - F_z(\text{inlet})(X_1)$$

$$\sum M_z = M_z(\text{inlet}) + M_z(\text{exhaust}) - F_x(\text{inlet})(Y_1) + F_y(\text{inlet})(X_1)$$

Example Problem

Check if the given actual loads at the inlet and exhaust nozzle of a single-stage vertically split steam turbine is below NEMA allowables. The inlet diameter is 3 in. and outlet diameter is 8 in. The NEMA coordinate system (X-axis parallel to the turbine shaft) is given in Figure 8.1. Two views of the turbine are given in Figures 8.2 and 8.3 (reference 3).

The orientation of X-, Y-, and Z-axes and the distances X_1, Y_1, Z_1 for the Example Problem are shown in Figures 8.2 and 8.3. The distances are measured from the centerline of the exhaust nozzle. The minus sign shown with X_1 and Z_1 distances corresponds with moment summations from Eq. 8.1. The sign for these distances depends upon the location of the inlet nozzle with respect to the exhaust nozzle in the NEMA system. Local forces and

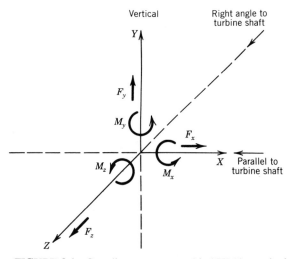

FIGURE 8.1   Coordinate system used in NEMA standard.

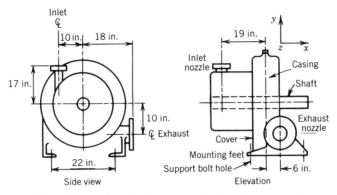

FIGURE 8.2 Typical single-stage vertically split steam turbine.

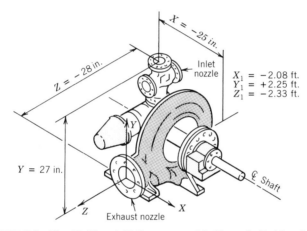

FIGURE 8.3 The X, Y, and Z distances used in Example Problem for Eq. 8.1.

TABLE 8.1 Forces and Moments from Analysis

Forces and Moments	Inlet	Exhaust
F_x (lb)	−30	−155
F_y	−55	1095
F_z	204	170
$F_{result.}$	213	1119
M_x (ft.-lb)	120	44
M_y	−67	−425
M_z	127	−722
$M_{result.}$	185	839

moments at the inlet and exhaust nozzles obtained from pipe stress analysis are listed in Table 8.1.

The components of resultant forces and moments after being transferred to the exhaust can be obtained by using Eq. 8.1.

$$F_x = (-30) + (-155) = -185 \text{ lb}$$

$$F_y = -55 + 1095 = 1040$$

$$F_z = 204 + 170 = 374$$

$$M_x = 120 + 44 - (-55)(-2.3) + 204(2.25) = 494.85 \text{ ft.-lb}$$

$$M_y = -67 + 425 + (-30)(2.33) - (204)(-2.08) = 2.22$$

$$M_z = 124 - 722 - (-30)(2.25) + (-55)(-2.08) = -416$$

The combined resultant force and moment after being transferred to the exhaust is as follows:

Combined resultant force at exhaust $= \sqrt{(-185)^2 + 1040^2 + 374^2} = 1121 \text{ lb}$

Combined resultant moment at exhaust $= \sqrt{494.85^2 + 2.22^2 + (-416)^2} = 647 \text{ ft-lb}$

Allowable Local Forces and Moments

The NEMA rule 1 applies for calculating allowable resultant local force:

for exhaust

$$F_{\text{allow}} = 166.6D - \frac{M}{3} = 166.6(8) - \frac{839}{3} = 1053 \text{ lb} \qquad \text{(Rule 1)}$$

A graph (Figure 8.4) can be used to determine the allowable resultant force. The calculation is shown by dotted lines.

For the inlet,

$$F_{\text{allow}} = 166.6(3) - \frac{185}{3} = 438 \text{ lb}$$

The allowable components of resultant forces and moments after being transferred to the exhaust is, using NEMA rule 2b,

$$\text{Equivalent diameter} = \sqrt{\frac{4}{\pi} \text{(total area of openings)}}$$

$$= \sqrt{D_{\text{inlet}}^2 + D_{\text{exhaust}}^2} = \sqrt{3^2 + 8^2} = 8.544$$

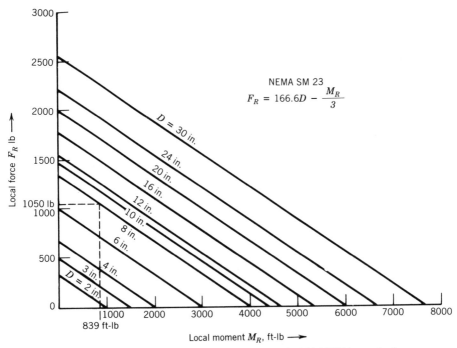

FIGURE 8.4 Force moment relationship using SM 23 NEMA standard.

This is below the 9 in. limit for the diameter given in the NEMA code. Therefore D_c = equivalent diameter = 8.544.

$$F_x = 50(8.544) = 427 \text{ lb} \qquad M_x = 250(8.544) = 2136 \text{ ft.-lb}$$

$$F_y = 125(8.544) = 1068 \text{ lb} \qquad M_y = 125(8.544) = 1068 \text{ ft-lb} \quad \text{(Rule 2b)}$$

$$F_z = 100(8.544) = 854.4 \text{ lb} \qquad M_z = 125(8.544) = 1068 \text{ ft-lb}$$

The allowable combined resultant force and moment at exhaust is, using NEMA Eq. 2a,

$$F_c = \frac{250D_c - M_c}{2} = \frac{250(8.544) - 647}{2}$$

$$= 744.5 \text{ lb} \qquad \text{(Rule 2a)}$$

When the actual load is higher than the allowable, the turbine vendors may be contacted to get the higher loads approved. It is the experience of the stress engineers that the allowable values are conservative. It would be very helpful if NEMA publishes the basis and criteria of the equations given.

TABLE 8.2   Comparison of Actuals to Allowables

Nozzle	Actual	Allowable	Remarks
Inlet	$F_R = 213\,\text{lb}$	$F_R = 438$	OK
Exhaust	$F_R = 1119$	$F_R = 1054$	Exceeds
Components	$F_x = -185$	$F_x = 424$	OK
	$F_y = 1040$	$F_y = 1068$	OK
	$F_z = 374$	$F_z = 854$	OK
	$M_x = 495$	$M_x = 2136$	OK
	$M_y = 2$	$M_y = 1068$	OK
	$M_z = -416$	$M_z = 1068$	OK
Combined resultant	$F_c = 1121$	$F_c = 745$	Exceeds

PIPING CONNECTED TO CENTRIFUGAL COMPRESSORS

American Petroleum Institute Standard 617 refers to NEMA SM 23 as the basis for allowable loads. The 1979 and 1973 editions have slightly different wordings in their texts.

API 617, 1979, Section 2.5.1, page 7, "Compressors shall be designed to withstand external forces and moments at least equal to 1.85 times the values calculated in accordance with NEMA SM 23."

API 617, 1973, Section 2.4.1, page 5, "Compressors shall be designed to withstand external forces and moments at least equal to a value calculated from the NEMA 23 formulas. For these calculations constants in the formulas shall be increased by a factor of 1.85."

The example presented earlier for steam turbines can be used the same way for centrifugal compressors except for the factor of 1.85. References 4 to 6 apply to centrifugal compressors. Special consideration for dynamic vibration is needed in the case of reciprocating compressors, which is outside the scope of the discussion here.

PIPING CONNECTED TO CENTRIFUGAL PUMPS

The API 610 (reference 7) standard gives equations to calculate allowable forces and moments in the case of centrifugal pumps for general refinery service. The criteria apply for pumps with 4 in. discharge nozzles or smaller (suction nozzles may be larger) and situations where the pump is constructed of steel or alloy steel. The modulus of elasticity of the piping material at operating temperature (known as hot modulus) can be used to calculate actual loads. Using hot modulus will result in lower loads because the piping is more flexible at higher temperature.

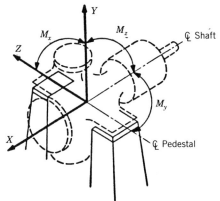

FIGURE 8.5 Coordinate system used for API 610 standard pumps.

Figure 8.5 gives the coordinate system used for API 610 pumps. The pump shaft is parallel to the X-axis and the Z-axis is along the centerline of the pedestal. For pumps with larger diameters, allowable values can be obtained from vendors or determined by experimental means (references 8 and 9).

The allowable force for each nozzle is:

$$F_x \leq 1.3\,W \leq 160\ \text{lb/nominal diameter}$$

$$F_z \leq 1.0\,W \leq 130\ \text{lb/nominal diameter}$$

$$F_y(\text{comp.}) \leq 200\ \text{lb/nominal diameter} \leq 1.2\,W$$

$$F_y(\text{tension} \leq 100\ \text{lb/nominal diameter} \leq 0.5\,W$$

where $W = $ weight of the pump.

The shear resultant force is:

$$F_r = \sqrt{F_x^2 + F_z^2} \leq 2000\ \text{lb}$$

The allowable moment is:

$$\sum M_x = 3.4\,W\ \text{ft-lb}$$

$$\sum M_y = 2.0\,W$$

$$\sum M_z = 1.5\,W$$

where the minimum value for W is 1000 lb.

The actual forces and moments from both suction and discharge nozzles shall be transferred to the intersection of X-, Y-, and Z-axes to obtain the summation of moments in each direction for comparison with allowables. The lesser of the values obtained from the considered weight and diameter should be used as the allowables.

PIPING YIELD METHOD

The piping yield method is an extreme case in which the component is given sufficient strength to fully yield the connecting piping in bending at the nozzle. The coordinate system and equations given in reference 10 are presented in Figure 8.6.

The forces and moments for which equipment needs to be designed are as follows:

$$F_{axial} = F_y = \frac{A_p}{2}\left(S_y - \frac{PD_I}{4t_p}\right) \qquad \text{axial force (along nozzle axis)}$$

$$M_x = M_z = 1.3Z_p\left(S_y - \frac{PD_I}{4t_p}\right) \qquad \text{bending moment}$$

$$M_{tor} = M_y = Z_p\sqrt{S_y^2 - \left(\frac{PD_I}{4t_p}\right)^2} \qquad \text{torsional moment}$$

Equations that are slightly different from these are also used in the industry (reference 11):

$$F_x = F_y = F_z = 0.01S_y \quad \text{(metal area of pipe)}$$

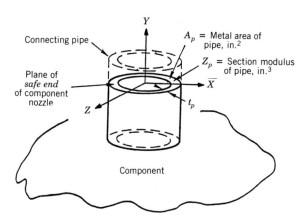

FIGURE 8.6 Pipe nozzle coordinate system.

The bending moment about the orthogonal directions (other than torsion) is:

$$M_x = M_z = 0.0707 S_y Z_p$$

The torsional moment is expressed by:

$$M_T = M_y = 0.1 S_y Z_p$$

The coordinate system for the above equations is the same as in Figure 8.6.

S_y = yield stress of pipe material   (maximum of 36,000 psi)

Z_p = section modulus of pipe, in.^{3}

Six Component Fraction Methods for Nozzle Loads Check

Six components fraction method requires that the sum of the ratios of the actual loads to vendor allowable loads be equal to or less than unity.

$$\frac{F_x}{F_{x\ allow}} + \frac{F_y}{F_{y\ allow}} + \frac{F_z}{F_{z\ allow}} + \frac{M_x}{M_{x\ allow}} + \frac{M_y}{M_{y\ allow}} + \frac{M_z}{M_{z\ allow}} \leq 1$$

REFERENCES

1. National Electrical Manufacturers Association. Publication No. SM 23, 1979 Sec. SM 23, 8.06, "Steam Turbine for Mechanical Drive Service."
2. General Electric Company. "Design Recommendations for Steam Piping Systems," Manual Number GEK-27,060.
3. Kannappan, S. et al. "How to Determine Allowable Steam Turbine Loads," *Hydrocarbon Processing*, Vol. 53, No. 8 (August 1974), p. 75.
4. American Petroleum Institute API 617, 4th ed. (November 1979), Sec. 2.5.1, "Centrifugal Compressors for Refining Service."
5. American Petroleum Institute API 617, 3rd ed. (October 1973), Sec. 2.4.1, "Centrifugal Compressors for Refinery Service."
6. Kannapan, S. "Determining Centrifugal Compressor Piping Loads," *Hydrocarbon Processing* (February 1982), p. 91.
7. American Petroleum Institute API 610. "Centrifugal Pumps for General Refinery Service," Sec. 14.
8. Simmon C. A. "Allowable Pump Piping Loads," *Hydrocarbon Processing* (June 1972), p. 98.
9. Doolin J. H. "Install Pumps for Minimum Stress," *Hydrocarbon Processing* (June 1978), p. 96.
10. Meyer R. A. "Survey of Nozzle Piping Reaction Criteria for Mechanical Equipment," *Structural Design of Nuclear Plant Facilities*, Vol. II, p. 283.
11. *Tenneessee Valley Authority.* "Allowable Loads for Equipments."

CHAPTER NINE

SPECIAL TOPICS

The topics that did not fall into the major categories of other chapters are grouped and discussed as follows:

Valves
Pressure relief valve thrust
Aluminum, Nickel, and Copper Alloy Piping
Underground and Plastic Piping
External Pressure Design—Jacketed Piping
Metric Units
Elevated Temperature—Creep Effects
Refractory Lining

VALVES

Valves are used in a piping system to achieve the following:

1. To stop or start flow of fluids. Examples are gate, plug (cocks), ball, or butterfly valves.
2. To regulate flow. Examples are globe, angle, needle, and butterfly valves.
3. To prevent back flow. Examples are lift check and swing check valves.
4. To regulate pressure. An example is regulators.
5. To relieve pressure. Examples are spring-loaded safety or pop valves, rupture disk relief valve.

There are numerous valve manufacturers making valves for many different uses. Reference 1 lists some of the manufacturers' information about valves for piping applications.

Gate Valves

Gate valves connect three major components: body, bonnet, and trim. (reference 1). The body is generally connected to the piping by means of flanged, screwed, or welded connections. The bonnet, containing the moving parts, is joined to the body generally with bolts to permit cleaning and maintenance. The valve trim includes the stem, the gate, the wedge or disc, and seat rings.

Valve Body Materials

Valve bodies are made of brass or bronze mainly in the smaller sizes and for moderate pressures and temperatures. Cast iron is used in most services. Cast steels are used for severe services of high pressures and high temperatures.

Valve Trim Materials

Valve trim materials include the seat ring, disc or facing, and stem. Common trim materials are monel, bronze, stellite, and stainless steel. Among the principal factors that influence the performance of trim materials are (1) tensile properties, chemical stability, and corrosion resistance at the operating temperature; (2) hardness and toughness; (3) a coefficient of expansion that corresponds closely to that of the valve body; and (4) difference in properties of seat and disc to prevent siezing.

The type of valve to be used for a given service is presented in the piping specification (Fig. 1.3). In general, preliminary stress analysis is carried out with approximate weight and actual weight obtained from the manufacturer (reference 1) should be used in the final stress analysis in critical systems. In nuclear piping, valves are further grouped as (1) active valves and (2) nonactive valves that are based on their requirement of performance after earthquake event (Chapter 10). Modeling of valves in computer analysis is described in Chapter 10 (Fig. 10.19). Valves require rigid support close to the center of gravity. It is advisable to avoid supports on the valve operators. In general, maximum acceleration a valve can be subjected to is 3 g. If actual acceleration exceeds allowable, valve vedor needs to be contacted.

Pressure Relief Valves

*Design Pressure and Velocity for Open Discharge Installation Discharge Elbows and Vent Pipes**

There are several methods available to the designer for determining the design pressure and velocity in the discharge elbow and vent pipe. It is the

---

* ANSI/ASME B31.1 Power Piping Code.

responsibility of the designer to assure himself that the method used yields conservative results. A method for determining the design pressures and velocities in the discharge elbow and vent pipe for open discharge installation is shown below and illustrated in the sample problem.

First, calculate the design pressure and velocity for the discharge elbow.

1. Determine the pressure P_1 that exists at the discharge elbow outlet (Fig. 9.1):

$$P_1 = \frac{W}{A_1}\frac{(b-1)}{b}\sqrt{\frac{2(h_o-a)J}{g_c(2b-1)}} \qquad (9.1)$$

2. Determine the velocity V_1 that exists at the discharge elbow outlet (Fig. 9.1):

$$V_1 = \sqrt{\frac{2g_cJ(h_o-a)}{(2b-1)}} \qquad (9.2)$$

where W = actual mass flow rate, lbm/sec
       A_1 = discharge elbow area, in.^{2}

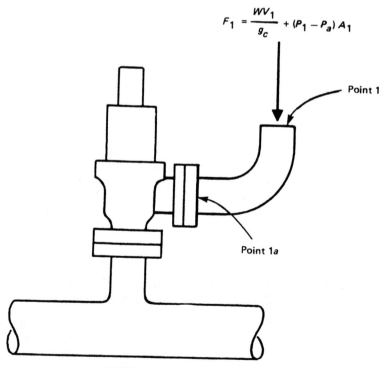

$$F_1 = \frac{WV_1}{g_c} + (P_1 - P_a)A_1$$

Point 1

Point 1a

FIGURE 9.1   Pressure relief valve.

h_o = stagnation enthalpy at the safety valve inlet, Btu/lbm
J = 778.16 ft-lbf/Btu
g_c = gravitational constant
    = 32.2 lbm-ft/lbf-sec^2
P_1 = pressure, psia (lbf/in.^{2}, absolute)
V_1 = ft/sec

Common values of a and b are listed in Table 9.1.

TABLE 9.1

Steam Condition	a, Btu/lbm	b
Wet steam, <90% quality	291	11
Saturated steam, ≥90% quality, 15 psia ≤ P_1 ≤ 1000 psia	823	4.33
Superheated steam, ≥90% quality, 1000 psia < P_1 ≤ 2000 psia^a	831	4.33

^{a}This method may be used as an approximation. For pressures >2000 psi, an alternate method should be used for verification.

Reaction Forces with Open Discharge Systems

Discharge Elbow

The reaction force F due to steady-state flow following the opening of the safety valve includes both momentum and pressure effects. The reaction force applied is shown in Figure 9.1, and may be computed by the following equation:

$$F_1 = \frac{W}{g_c} V_1 + (P_1 - P_a)A_1 \tag{9.3}$$

where F_1 = reaction force, lbf at point 1
    W = mass flow rate, (relieving capacity stamped on the valve ×1.11), lbm/sec
    g = gravitational constant
      = 32.2 lbm-ft/lbf-sec^2
    V_1 = exit velocity at point 1, ft/sec
    P_1 = static pressure at point 1, psia
    A_1 = exit flow area at point 1, in.^{2}
    P_a = atmospheric pressure, psia

To ensure consideration of the effects of the suddenly applied load F, a dynamic load factor DLF should be applied (Fig. 9.2).

The methods for calculating the velocities and pressures at the exit point of the discharge elbow are the same as those discussed in Eqs. 9.1 and 9.2.

ANALYSIS FOR REACTION FORCES DUE TO VALVE DISCHARGE

Open Discharge Systems

The moments due to valve reaction forces may be calculated by simply multiplying the force, calculated as described in Eq. 9.3, times the distance from the point in the piping system being analyzed, times a suitable dynamic load factor. In no case shall the reaction moment at the branch connection below the valve be taken at less than the product as given in Eq. 9.4

$$\text{Moment} \geq (DLF)(F_1)(D) \tag{9.4}$$

where F_1 = force calculated in Eq. 9.3

$\quad D$ = nominal OD of inlet pipe

$\quad DLF$ = dynamic load factor (Fig. 9.2)

Reaction force and resultant moment effects on the header, supports, and nozzles for each valve or combination of valves blowing shall be considered.

Dynamic Amplification of Reaction Forces

In a piping system acted upon by time varying loads, the internal forces and moments are generally greater than those produced under static application of the load. This amplification is often expressed as the dynamic load factor DLF and is defined as the maximum ratio of the dynamic deflection at any time to the deflection which would have resulted from the static application of the load. For structures having essentially one degree-of-freedom and a single load application, the DLF value will range between one and two depending on the time-history of the applied load and the natural frequency of the structure. If the run pipe is rigidly supported, the safety valve installation can be idealized as a one degree-of-freedom system and the time-history of the applied loads can often be assumed to be a single ramp function between the no-load and steady-state condition. In this case the DLF may be determined in the following manner.

1. Calculate the safety valve installation period T using the following equation and Figure 9.2:

$$T = 0.1846 \sqrt{\frac{Wh^3}{EI}} \qquad (9.5)$$

where T = safety valve installation period, sec
   W = weight of safety valve, installation piping, flanges, attachments, etc., lb
   h = distance from run pipe to centerline of outlet piping, in.
   E = Young's modulus of inlet pipe, lb/in.^{2}, at design temperature
   I = moment of inertia of inlet pipe, in.^{4}

2. Calculate ratio of safety valve opening time to installation period (t_o/T) where t_o is the time the safety valve takes to go from fully closed to fully open (sec) and T is determined in (1) above.

3. Enter Figure 9.2 with the ratio of safety valve opening time to installation period and read the DLF from the ordinate. The DLF shall never be taken less than 1.1.

If a less conservative DLF is used, the DLF shall be determined by calculation or test.

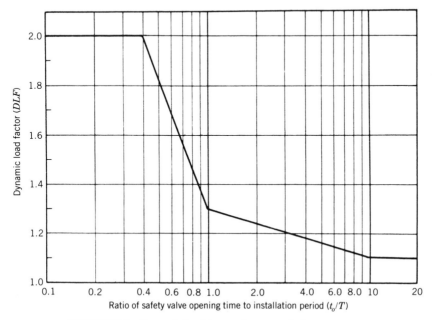

FIGURE 9.2 Dynamic load factors for open discharge system.

Example

Reaction Forces with Open Discharge Systems Calculation

Calculate reaction force with the following (see Fig. 9.3):

Operating temperature = 700°F

Operating pressure   = 200 psig

h_o = 1374 Btu/lbm (from steam table)

a = 823 Btu/lbm b = 4.33 (from Table 9.1)

$$W = \frac{20{,}000}{3600} = 55.6 \text{ lbm/sec}$$

A_1 = 50 in.^{2}

J = 778.16 ft-lbf/Btu

g_c = 32.2 lbm-ft/lbf-sec^2

Assume dynamic load factor of 2.0.

Therefore

$$P_1 = \frac{W}{A_1} \times \frac{(b-1)}{b} \times \sqrt{\frac{2(h_o - a) \times J}{g_c(2b-1)}}$$

$$= \frac{55.6}{50} \times \frac{(4.33-1)}{4.33} \times \sqrt{\frac{2(1374 - 823) \times 778.16}{32.2(2 \times 4.33 - 1)}}$$

$$= 50 \text{ psig}$$

$$V_1 = \sqrt{\frac{2g_c J(h_o - a)}{(2b-1)}} = \sqrt{\frac{2(32.2)(778.16)(1374 - 823)}{(2 \times 4.33 - 1)}}$$

$$= \sqrt{\frac{50{,}113.5(1374 - 823)}{(2 \times 4.33 - 1)}}$$

$$= 1899 \text{ ft/sec}$$

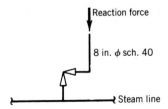

Reaction force

8 in. ϕ sch. 40

Steam line

FIGURE 9.3 Example Problem for open discharge system.

Therefore

$$F_1 = \frac{W}{g_c} V_1 + (P_1 - P_a)A_1$$

$$= \frac{27.8}{32.3} \times 1899 + (50 - 14.7)50$$

$$= 3404.5$$

$$F_{dyn} = F(DLF) = 3404.5(2) = 6809 \text{ lb}$$

ALUMINUM PIPING

Different aluminum alloy piping has similar desirable corrosion resistance but varies in mechanical properties. Aluminum alloys most commonly used for piping systems are alloy 160, alloy 3003, alloy 5052, alloy 6061, and alloy 6063. Of these, alloy ASTM B241-6063-T6 is the most widely used because it has good mechanical properties at reasonable cost.

Aluminum has found various uses in the cryogenic or cold temperature applications. As temperature decreases aluminum shows increased values of tensile and yields strength with equal or improved ductility or impact

TABLE 9.2   Physical Properties of Various Aluminum Alloys

	Alloy			
	3003	5052	6061	6063
Tensile strength, psi	17,000	41,000	45,000	35,000
Yield strength, psi	8,000	36,000	40,000	31,000
Modulus of elasticity, psi $\times 10^6$	10	10.2	10	10
Thermal conductivity, Btu/hr/sq ft/°F/in.	1,070	960	900	1,090
Average coefficient of thermal expansion, in./°F/in. $\times 10^{-6}$				
−58 to 68°	12.0		2.1	12.1
68 to 212°	12.9	13.2	13.0	13.0
68 to 392°	13.5		13.5	13.6
68 to 572°	13.9		14.1	14.2

TABLE 9.3   Support Spacing for Aluminum Alloy 6063 Pipe

Nominal Pipe Size (in.)	Pipe Schedule Number		
	5 S	10 S	40 S
	Support Spacing (ft)		
$\frac{1}{2}$	5.0	5.5	6.0
$\frac{3}{4}$	5.5	6.0	6.5
1	6.0	7.0	7.5
$1\frac{1}{2}$	6.5	7.5	8.5
2	6.5	8.0	9.0
3	7.5	8.5	10.0
4	8.0	9.0	11.5
6	9.0	10.0	13.0
8	9.5	11.0	14.0
10	10.0	11.5	15.0
12	10.5	13.0	15.5

resistance. The specific alloys most frequently used in cold temperature applications are alloy 3003 and alloy 5052.

Aluminum alloys may be welded. The inert–gas tungsten–arc method, using argon gas, is the recommended procedure. Further reference material about installation techniques, fittings, and so on can be obtained from manufacturers such as Alcoa and Reynolds metal companies.

Table 9.2 shows the physical properties of various aluminum alloys (reference 4). As can be seen, units for average thermal expansion is given as in./in./°F. See Chapter 1 (Eq. 1.1) for conversion to in./linear feet of pipe. The thermal expansion for aluminum is high and adequate provision must be made to compensate for the high amount of expansion. The operating pressure of an aluminum pipe is calculated in the same way as ferrous pipe, using Eq. 2.4 with $y = 0.4$ and corrosion allowance of zero.

Aluminum is subject to galvanic corrosion in the presence of an electrolyte. When in the presence of carbon steel, copper, brass, nickel, monel, tin, and lead, aluminum will be corroded. Thus conventional carbon steel pipe hangers should be avoided. However, the 300 series of stainless steel and zinc are usually compatible with aluminum. Therefore galvanized (zinc coating) steel hangers, aluminum hangers, or padded hangers may be used. Table 9.3 gives support spacing (reference 4) for alloy 6063 piping. This table is based on the pipe being uninsulated, operating at a maximum temperature of 400°F, and conveying a liquid of specific gravity 1.35 (a conservative assumption). If the line is insulated, reduce the span by 30%. No allowance has been made for concentrated loads such as valves. Equations 3.1 and 3.2 may also be used to calculate the span.

COPPER ALLOY PIPE

Many applications of the copper pipe have been found in the food industry. Table 9.4 shows the physical properties. The coefficient of thermal expansion of copper pipes is high. Therefore either loops or expansion joints must be provided for absorbing the expansion. Copper pipe can be joined by threading, soldering, or brazing, and flanges may be installed by any of these methods. Table 9.5a gives dimensions of the copper pipe (suitable for threading), outside diameter, wall thickness, and maximum allowable pressure at a operating temperature of 300°F for regular schedule. A copper pipe should not be forced into place during installation. Forcing the pipe into place and keeping it under stress can cause failure. Table 9.5b gives support spacing for a copper pipe. The spacing is based (reference 4) on uninsulated lines operating at a maximum temperature of 300°F and carrying a fluid of specific gravity 1.35. If the lines are insulated, spacing should be reduced by 30%. No allowance has been made for concentrated loads like valves. In order to prevent galvanic corrosion, copper or padded hangers should be used with copper piping. Table 9.6 shows allowable stress for alloys of nickel, copper, and aluminum for B31.3, B31.1, and section III codes.

A number of nonferrous metals and their alloys are wildly used as corrosion-resistant piping material. Zirconium and titanium are examples of

TABLE 9.4 Physical Properties of Copper Alloy

Tensile Strength	54,000 psi
Thermal conductivity at 68°F	2364 Btu/hr/sq ft/°F/in.
Average coefficient of linear thermal expansion 77–572°F	9.8×10^{-6} in./°F/in.
Modulus of elasticity in tension	17×10^6 psi

TABLE 9.5a Dimensions of Standard Copper Pipe (Suitable for Threading) and Maximum Allowable Operating Pressure (psi)

| Nominal Pipe Size | Outside Diam. | Regular Schedule | | |
		Inside Diam.	Wall Thickness	Allowable Pressure at 300°F or Lower
$\frac{1}{8}$	0.405	0.281	0.062	220
$\frac{1}{4}$	0.540	0.376	0.082	540
$\frac{3}{8}$	0.675	0.495	0.090	540
$\frac{1}{2}$	0.840	0.626	0.107	550
$\frac{3}{4}$	1.050	0.822	0.114	500
1	1.315	1.063	0.126	340
$1\frac{1}{4}$	1.660	1.368	0.146	430
$1\frac{1}{2}$	1.900	1.600	0.150	390
2	2.375	2.063	0.156	330
$2\frac{1}{2}$	2.875	2.501	0.187	270
3	3.500	3.062	0.219	310
$3\frac{1}{2}$	4.000	3.500	0.250	350
4	4.500	4.000	0.250	310
5	5.562	5.062	0.250	250
6	6.625	6.125	0.250	210
8	8.625	8.001	0.312	220
10	10.750	10.020	0.365	220
12	12.750	12.000	0.375	210

TABLE 9.5b Support Spacing for Copper Pipe for Regular Schedule (reference 4)

Nominal Pipe Size (in.)	Spacing, ft
$\frac{1}{2}$	6.5
1	8.0
$1\frac{1}{2}$	9.5
2	10.5
3	12.5
4	13.5
6	15.5
8	17.0
10	20.0

TABLE 9.6 Allowable Stress (ksi) for Nonferrous Alloys

Material	Code	Metal Temperature, °F													Yield Stress (ksi)
		−325 to 100°F	150	200	250	300	350	400	450	500	550	600	650	700	
Seamless Aluminum Alloy Pipe B241 6061 P No. 23 Temper T6	B31.3	12.7	12.7	12.7	12.3	10.5	7.9	5.6	–	–	–	–	–	–	35
	B31.1	9.5	9.5	9.5	9.1	7.9	6.3	4.5	–	–	–	–	–	–	–
	Sec. III Class Cl 3	9.5	9.5	9.5	9.1	7.9	6.3	4.4	–	–	–	–	–	–	35
Welded Copper Alloy B467 over 4½ in. φ Cu–Ni 90/10	B31.3	7.4	7.0	6.9	6.8	6.6	6.5	6.4	6.2	6.1	5.9	5.1	–	–	13
	B31.1	7.4	7.0	6.9	6.8	6.6	6.5	6.4	6.2	6.1	5.2	4.3	–	–	–
	Sec. III Class 2 and 3	8.7		8.1		7.8		7.5					–	–	13
Seamless Nickel Alloy B161 Alloy No. 200 P NO. 41 Annealed over 5 in. φ	B31.3	∞	∞	∞	∞	∞	∞	∞	∞	∞	∞	∞	–	–	12
	B31.1	∞	∞	∞	∞	∞	∞	∞	∞	∞	∞	∞	∞	–	–

TABLE 9.7   Properties of Corrosion-Resistant Metals and Alloys

Metal/Alloy	ASTM Spec.	Yield Stress (ksi)	Density (lb/in.^{3})	Specific Gravity	Thermal Exp. Coef. (in./in./°F)	E Modulus (psi)
Incoloy 800	B407	42	0.290	8.01	7.9×10^{-6}	28×10^6
Incoloy 825	B423	35	0.294	8.14	7.8×10^{-6}	28×10^6
Inconel 600	B167	36	0.304	8.43	7.4×10^{-6}	31×10^6
Zirconium Gr. 702 unalloyed			0.234			14.4×10^6
Titanium, Gr. 1	B337	25				
Hastelloy B2, ASME code case 1642	B517	51	0.333		6.2×10^{-6}	31.4×10^6
Hastelloy C276, ASME code case 1410		59.8	0.321		7.1×10^{-6}	29.8×10^6

these metals. Incoloy, hastelloy, and inconel are alloys that have good corrosion-resistant properties. These alloys are trade names. Their ASTM specification numbers and properties are given in Table 9.7. The ASTM annual 1977 or 1980, part 8, gives more information.

UNDERGROUND PIPING

Routing of piping underground is sometimes necessary to cross a road, piping between buildings, yard crossing, and so on. The factors that are important in underground plastic piping design are as follows:

1. Longitudinal bending stress
2. Buckling, arching
3. Bowing
4. Soil stiffness and soil geometry (very important for expansive soils)
5. Dead and live load
6. Wall compression, bending, and shear resistance
7. Hydrostatic uplift

Modes of failure and collapse are described below as applied to plastic piping (reference 8):

1. Caving due to deflection (see Fig. 9.4a)
2. Wall(ring) compression due to yielding at A (Fig. 9.4b)

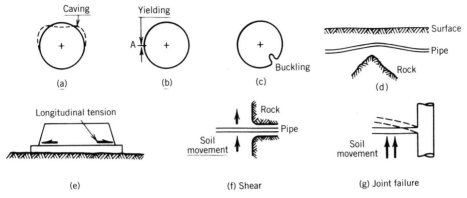

FIGURE 9.4   Underground piping failure modes.

3. Buckling: (a) elastic and (b) plastic. In buckled state a pipe resembles the one shown in Fig. 9.4c
4. Beam bending (Fig. 9.4d)
5. Longitudinal tension (along axis) (see Fig. 9.4e)
6. Direct shear (occurring at hard soft interfaces) (see Fig. 9.4f)
7. Failure at joint (See Fig. 9.4g)

Heat transfer loss from a buried pipeline has become more important in case of heated oil pipeline or in the case of underground steam pipes for keeping ice from sidewalks and driveways. Equations to calculate the heat transfer from a buried pipe is given in reference 3.

Underground Piping Design

Design the following underground line. Assume data as necessary. The depth of cover = H = 3 ft; trench width = 2 ft 6 in.; material is ASTM A53 Grade B; minimum specified yield stress is 35,000 psi at 145°F. See Figure 9.5. The OD of the pipe including insulation is 12 in.

Assume the coefficient of friction between pipe and soil is 0.3.
Density of saturated clay soil ω = 100 lb/ft^3.
Pipe contains #6 fuel oil 13 API with specific gravity 1.2.

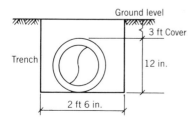

FIGURE 9.5   Buried pipe with trench dimensions.

Steps

1. Calculate frictional resistance F_f.
2. Calculate thermal force F_t after calculating longitudinal stress S_L.
3. Find point of no movement in which $F_f = F_t$.
4. Calculate hoop stress, bending stress due to earth load, radial stress due to pressure, and temperature stress due to operating temperature.
5. Calculate combined stress from stresses in (4) using maximum strain theory.
6. Find maximum allowable stress using B31.4 liquid transportation piping code.
7. For proper design the combined stress should be less than the maximum allowable stress.

First, calculate the load on the pipe from the backfill W_c.

$$W_c = \text{load on pipe, lb/ft}$$

$$W_c = C_d w B_d^2 . \tag{9.6}$$

where

$$C_d = \text{load coefficient}$$

$$= \frac{1 - e^{-2k\mu'(H/B_d)}}{2 k\mu'} \tag{9.7}$$

where k = ratio of lateral unit pressure to vertical unit pressure
μ' = coefficient of friction between fill material and ditch $\leq \mu$, where μ is the coefficient of internal friction of fill
H = height of fill above top of pipe, feet
B_d = horizontal width of ditch, feet

Values of the load coefficient C_d may be taken from the diagram in Figure 9.6 (reference 7).

Step 1. Calculation of Frictional Resistance (units lb/ft): Let W_c = load on pipe from backfill neglecting moving load.

Using Marston's formula:

$$W_c = C_d w B_d^2 \tag{9.6}$$

where C_d = load coefficient. Read from graph in Figure 9.6:
$W_c = 1.0 \times 100 \times 2.5 \times 2.5 = 625$ lb/ft
$C_d = 1$ for $\dfrac{H}{B_d} = \dfrac{3}{2.5} = 1.2$
w = density of soil = 100 lb/ft^3
B_d = trench width = 2.5

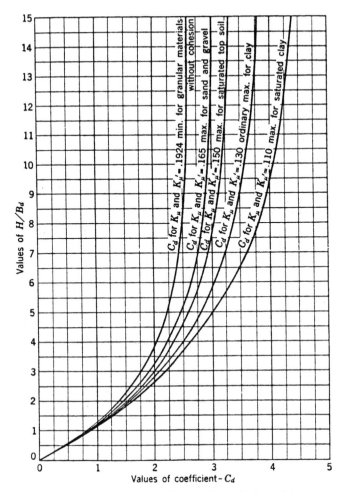

FIGURE 9.6 Load coefficient C_d for ditch conduits.

Weight of content = (weight of water) specific gravity
                            = (21.69)1.2 = 26 lb/ft
Insulation weight   = 5.38 lb/ft
Pipe metal weight   = 28.55 lb/ft
Total weight of pipe = 625 + 26 + 5.38 + 28.55 = 685 lb/ft
Frictional resistance = μ(total weight) = 0.3(685) = 205.5 lb/ft

Steps 2 and 3. Point of No Movement: Frictional force opposes thermal force. At the point of no movement the frictional force is equal to the thermal expansion force.

$$\text{Longitudinal stress} = S_L = E\alpha(T_2 - T_1) - \left(vP\left(\frac{D-2t}{2t}\right)\right) \quad (9.8)$$

where $E = 27.9 \times 10^6$ psi
    ν = Poisson's ratio = 0.3
    $D = 8.625$ in.
    α = linear coefficient of thermal expansion
    $T_2 = 145°F$
    $T_1 = 80°F$
    $t = 0.322$ in.
    $P = 300$ psig

Equation 9.8 is from B31.4 Piping Code (reference 3 in Chapter 4) Section 419.6.4(b) of the Liquid Transportation Piping Code. This equation is for restrained piping, in this case because of underground piping:

$$S_L = 27.9 \times 10^6 (6.5 \times 10^{-6})(145 - 80) - 0.3(300)\left[\frac{8.625 - 2(0.322)}{2(0.322)}\right]$$

$$= 11{,}787 - 0.3(3717) = 10{,}669 \text{ psi}$$

Thermal expansion force $F_t = S_L$ (metal area) $= 10{,}669(8.4) = 89{,}625$ lb

Distance of point of no movement from point of burial $= \dfrac{89{,}625}{205.5}$

$$= 436 \text{ ft}$$

Figure 9.7 shows the distance 436 ft from the point of burial marked.

Step 4. Stresses: *Hoop stress* (S_1):

$$S_1 = \frac{P(D - 2t)}{2t} = \frac{300(8.625 - 0.644)}{0.644} = 3717 \text{ psi}$$

For bending stress (S_2), use Spangler's equation:

$$S_2 = 0.177 \frac{(C_d w B_d^2) Et R_m}{Et^3 + 2.592 PR_m^3} \tag{9.9}$$

where R_m = mean radius of pipe $= (D - t)/2 = 4.1515$ in. Other terms are the same as for Eq. 9.8

$$S_2 = 0.177\left[\frac{625(27.9 \times 10^6)(0.322)(4.1515)}{27.9 \times 10^6 \times 0.322^3 + 2.592(300)(4.1515)^3}\right]$$

$$= 4179 \text{ psi}$$

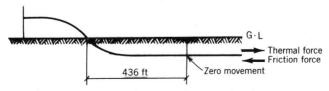

FIGURE 9.7 Point of zero movement in a buried pipe.

Radial stress (S_3):

$$S_3 = P = 300 \text{ psi}$$

Thermal stress (S_4):

$$S_4 = E\alpha(T_2 - T_1) = 27.9 \times 10^6 \times 6.5 \times 10^{-6}(145 - 80)$$
$$= 11{,}787.73 \text{ psi}$$

S_c = *circumferential stress*

$\quad$ = $S_{\text{hoop}} + \nu$ (thermal stress + bending stress due to earth)

$\quad$ = $3717.85 + 0.3(11{,}787.73 + 4179)$

$\quad$ = 8507 psi

Step 5. Combined stress (S): Use maximum strain theory:

$$S = \sqrt{S_L^2 + S_c^2 + S_R^2 - 2\gamma(S_L S_c + S_L S_R + S_c S_R)}$$
$$= \sqrt{10{,}669^2 + 8507^2 + 300^2 - 2(0.3)(10{,}670 \times 8507 + 300(10{,}670 + 8507)}$$
$$= 11{,}327 \text{ psi}$$

Step 6. Maximum allowable stress: Use restrained piping according to B31.4:

$$\text{Maximum allowable stress} = 0.90(\text{minimum } \sigma_y \text{ of pipe})$$
$$= 0.90(35{,}000)$$
$$= 31{,}500 \text{ psi}$$

Step 7: The combined stress of 11,327 psi is below allowable stress of 31,500 psi. Thus the pipe design is safe.

EXTERNAL PRESSURE DESIGN

The design of cylindrical vessels that is subjected to external pressure is outlined in the ASME Section VIII, Division 1 UG28(c).

Nomenclature:
P_c = critical collapsing pressure, psi
P_a = allowable pressure, psi
$\quad t$ = wall thickness, inches
E = modulus of elasticity, psi
L = length between stiffners, inches

L_c = critical length, inches
S_y = yield stress, psi
S_c = tangential stress at collapse pressure, psi
Factor $A = S_c/E$
Factor $B = S_c/2$

Strength of Pipe Under External Pressure (reference 10)

The strength of pipe under external pressure is a function of the physical properties of the construction material at the operating temperature and its geometrical parameters such as the unsupported length L, pipe thickness t, the outside diameter D_o, and the pipe out-of-roundness.

The behavior of thin-wall cylindrical shells under uniform external pressure varies according to cylinder length as follows:

1. *Very Long Cylinders*: The critical collapsing pressure is given by:

$$P_c = 2.2E\left(\frac{t}{D_o}\right)^3 \tag{9.10}$$

The critical length L_c, the minimum unsupported length beyond which P_c is independent of L, is given by:

$$L_c = 1.11D_o\left(\frac{D_o}{t}\right)^{1/2} \tag{9.11}$$

2. *Intermediate Cylinders with $L < L_c$*: The critical pressure P_c is a complicated function of the collapsed contour and the two characteristic ratios t/D_o and L/D_o. For practical design, P_c can be given by the following empirical equation:

$$P_c = \frac{2.8E(t/D_o)^{2.5}}{L/D_o} \tag{9.12}$$

3. *Short Cylinders*: The cylinder will fail in this case by plastic yielding. The critical pressure can be given by:

$$P_c = \frac{2S_y t}{D_o} \tag{9.13}$$

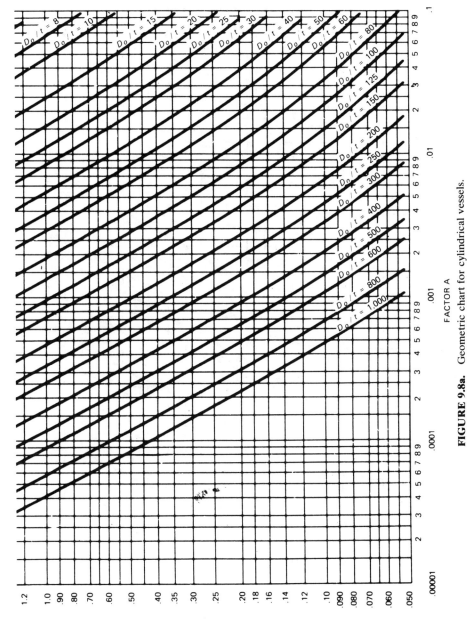

FACTOR A

FIGURE 9.8a. Geometric chart for cylindrical vessels.

151

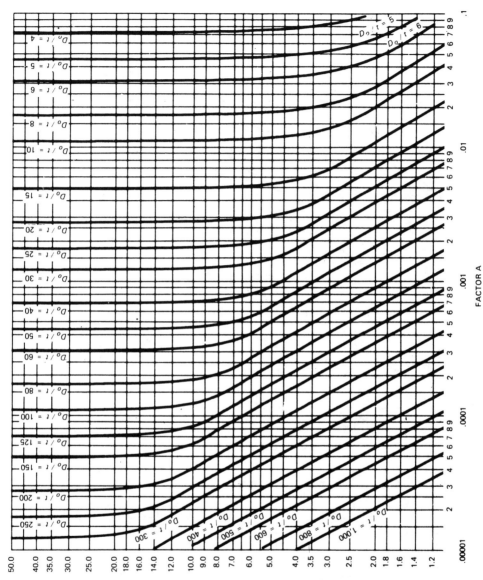

FACTOR A

FIGURE 9.8b. Geometric chart for cylindrical vessels.

Length ÷ Outside Diameter = L/D_o

152

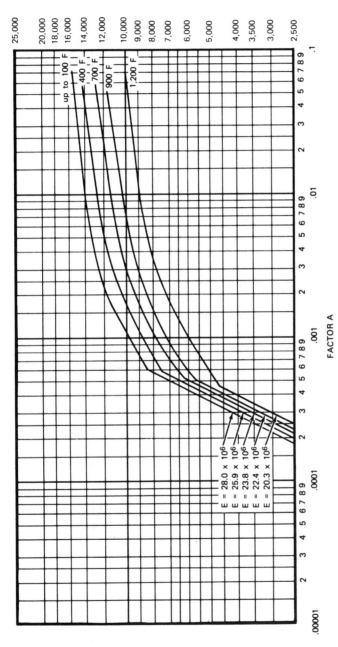

FIGURE 9.9a. Chart for determining shell thickness of cylindrical and spherical vessels under external pressure when constructed of austenitic steel [18 Cr–8 Ni–Mo, Type 316; 18 Cr–8 Ni–Ti, Type 321; 18 Cr–8 Ni–Cb, Type 347, 25 Cr–12 Ni, Type 309 (through 1100°F only); 25 Cr–20 Ni, Type 310; and 17 Cr, Type 430B stainless steel (through 700°F only)¹] (ASME Section VIII, Division 1).

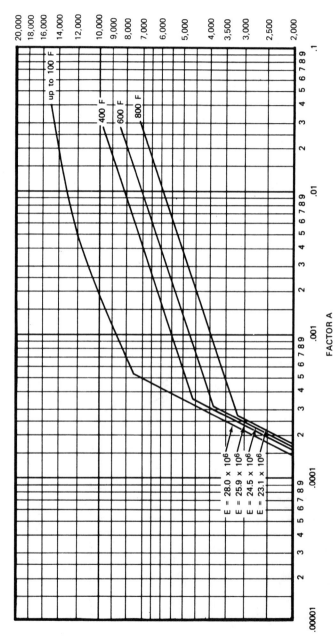

FIGURE 9.9b. Chart for determining shell thickness of cylindrical and spherical vessels under external pressure when constructed of austenitic steel (18 Cr–8 Ni–0.03 maximum carbon, Type 304L) (ASME Section VIII, Division 1).

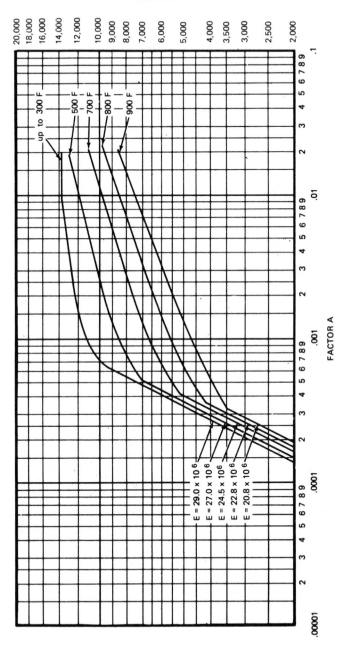

FIGURE 9.10a. Chart for determining shell thickness of cylindrical and spherical vessels under external pressure when constructed of carbon or low-alloy steels (specified yield strength 24,000 psi to, but not including, 30,000 psi) (ASME Section VIII, Division 1).

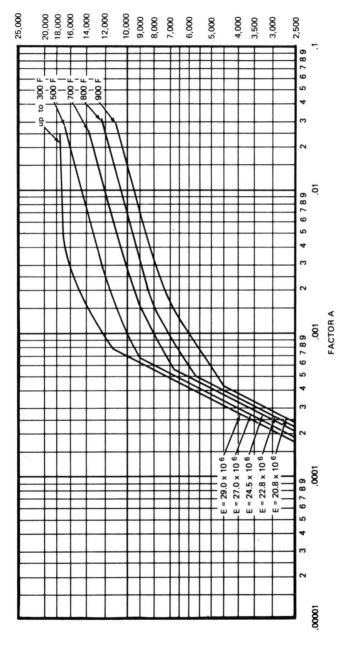

FACTOR B

FACTOR A

up to 300 F
500 F
700 F
800 F
900 F

E = 29.0 x 10⁶
E = 27.0 x 10⁶
E = 24.5 x 10⁶
E = 22.8 x 10⁶
E = 20.8 x 10⁶

25,000
20,000
18,000
16,000
14,000
12,000
10,000
9,000
8,000
7,000
6,000
5,000
4,000
3,500
3,000
2,500

FIGURE 9.10b. Chart for determining shell thickness of cylindrical and spherical vessels under external pressure when constructed of carbon or low-alloy steels [specified minimum yield strength 30,000 psi (207 MPA) and over] and Type 405 and Type 410 stainless steels (ASME Section VIII, Division 1).

156

ASME Charts

Geometric Chart: For $L > L_c$, the tangential stress S_c at collapse pressure P_c (given by Eq. 9.10), can be written as:

$$S_c = \frac{P_c D_o}{2t} = 1.1E\left(\frac{t}{D}\right)^2 \tag{9.14}$$

$$\text{Strain} = \frac{S_c}{E} = 1.1\left(\frac{t}{D_o}\right)^2 \tag{9.15}$$

For $L \leq L_c$, the tangential stress S_c at collapse pressure P_c (given by Eq. 9.13), is:

$$S_c = \frac{\left[1.4E\left(\dfrac{t}{D_o}\right)^{1.5}\right]}{\dfrac{L}{D_o}} \tag{9.16}$$

$$\text{Strain} = \frac{S_c}{E} = \frac{1.4\left(\dfrac{t}{D_o}\right)^{1.5}}{\dfrac{L}{D_o}} \tag{9.17}$$

Equations 9.14 and 9.16 were plotted to develop the geometric chart in Figure UGO-28.0 of the ASME Boiler and Pressure Vessel Code, Section VIII, Division 1. (See Fig. 9.8a and Fig. 9.8b.)

Material Chart: These charts are actually stress–strain curves for materials at design temperatures (S_c/E on the abscissa and $S_c/2$ values as variables on the ordinate). (S_c/E is called factor A and $S_c/2$ is called factor B in ASME notation.)

The allowable pressure can be obtained by using the following relation:

$$\frac{S_c}{2} = B = \frac{P_c D_o}{4t} \quad \text{or} \quad P_c = \frac{4Bt}{D_o}$$

Let $P_c = 3P_a$(safety factor = 3) $P_a = \frac{4}{3}B\left(\frac{t}{D_o}\right)$ \qquad (9.18)

Equation 9.18 is the same as Section VIII subsection UGO-28(c)-1 equation.

Practical Design Using ASME Charts: Determine the pipe thickness t under external pressure ($D/t \geq 10$).

Step 1.   Assume a value for t and determine the ratios L/D and D/t.

Step 2.   Determine the value of factor A from the geometric chart (using values obtained in Step 1).

Step 3.   Determine the value of factor B by using the proper material chart and the value of factor A obtained in Step 2.

Step 4.   Calculate the allowable external pressure P_a by using Eq. 9.18.

Step 5.   For values of A falling to the left of the applicable material/temperature line. The value of P_a can be determined by the following formula:

$$P_a = \frac{2}{3} AE \frac{t}{D_o} \tag{9.19}$$

(*Note*:   For $D/t < 10$ use the procedure outlined in ASME UG-28(c)-2.)

Step 1.   Assume a value for t and calculate B:

$$B = \frac{3PD_o}{4t}$$

Step 2.   Find the factor A by using the proper material chart and the value of B obtained above. If the value for factor B is less than the value listed in the chart, factor A is given by:

$$A = \frac{1.5PD_o}{Et}$$

Step 3.   Determine the value of L/D by entering the factor A and the appropriate D/t curve in the geometric chart. The maximum unstiffened length is obtained by multiplying the value of L/D by D. (If there is no intersection between the vertical projection of A and the D/t curve, then stiffeners are not required for any length.)

See ASME Section VIII, Division 1, subsection UGO-29 for design of stiffening ring design.

Example (reprinted from ASME Appendix L, Section VIII, Division 1)

Vessels Under External Pressure

Note: The lines on Figure UGO-28.0 of Appendix V (Fig. 9.8a and b) express a geometrical relationship between L/D_o and D_o/t for cylindrical shells and tubes which is common for all materials. This chart is used only for determining the factor A when factor A is not obtained by formula in the special case when $D_o/t < 10$. (See UG-28(c)-2.)

The remaining charts in Appendix V are for specific material or classes of materials and represent pseudo stress–strain diagrams containing suitable factors of safety relative both to plastic flow and elastic collapse.

Reference from ASME Section VIII, UA-270

(a) *Cylindrical Shell Under External Pressure.* [An example of the use of the rules in UG-28(c)] [Eq. 9.18].

GIVEN: Fractionating tower 14 ft inside diameter by 21 ft long, bend line to bend line, fitted with fractionating trays, and operating under a vacuum at 700°F. The tower to be constructed of SA-285 Grade C carbon steel. Design length is 39 in.

REQUIRED: shell thickness, t

SOLUTION:

Step 1.
Assume a thickness $t = 0.3125$ in.
Assumed outside diameter $D_o = 168.625$ in.

$$\frac{L}{D_o} = \frac{39}{168.625} = 0.231$$

$$\frac{D_o}{t} = \frac{168.625}{0.3125} = 540$$

Steps 2, 3. Enter Fig. UGO-28.0 in Appendix V (Fig. 9.8a and Fig. 9.8b) at the value of $L/D_o = 0.231$; move horizontally to the D_o/t line of 540 and read the value A of 0.0005.

Steps 4, 5. Enter Fig. UCS-28.2 (Fig. 9.10b) at the value of $A = 0.0005$ and move vertically to the material line for 700°F. Move horizontally and read B value of 6100 on ordinate.

Step 6. The maximum allowable external pressure [Eq. 9.18] for the assumed shell thicknesses of 0.3125 in. is:

$$P_a = \frac{4B}{3(D_o/t)} = \frac{4(6100)}{3(540)} = 15.1 \text{ psi}$$

Since P_a is greater than the design pressure P of 15 psi, the assumed thickness should be satisfactory.

JACKETED PRESSURE PIPING SYSTEM

Method of Calculating Core Pipe Thickness

Example 1:   The pipe is 4 in., operating under a full vacuum at 650°F. The external pressure is 100 psig. The maximum length of spool piece without stiffening rings is 10 ft 0 in. Pipe material is stainless steel ASTM A-312, Type 316 (Fig. 9.11).

Calculate the core pipe thickness.

Step 1.   Assume the thickness of pipe is sch 10 S (0.120 in.).

T_1(min thickness) $= 0.120 \times 0.875$ (assume 12.5% manufacturing
tolerance)

$$= 0.105 \text{ in.}$$

D_o(outside dia.) $= 4.50$ in.

l(spool length) $= 120$ in.

thus

$$\frac{L}{D_o} = \frac{120}{4.50} = 26.66$$

$$\frac{D_o}{t} = \frac{4.50}{0.105} = 42.85$$

Step 2.   Enter value of $L/D_o = 26.66$ (Fig. 9.8b) in Figure UGO-28.0, Appendix V, ASME Section VIII, Division 1, move horizontally to the D_o/t line of 42.85 and read the value A of 0.0006.

For value of $L/D_o > 50$, enter the chart at 50 for determining the value of constant A.

Step 3.   Enter (Fig. 9.9a) in Figure UHA-28.2 at the value of $A = 0.0006$ and move vertically to the material line for 650°F (interpolated between the 400 and 700°F material lines). Move horizontally and read B value of 6700 on the ordinate.

Step 4.   The maximum allowable external pressure (Eq. 9.18) for the

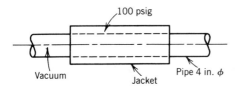

Vacuum   Jacket   Pipe 4 in. ϕ

FIGURE 9.11   Jacketed piping (vacuum in the process pipe).

assumed sch 10 S core pipe is:

$$P_a = \frac{4}{3} B \frac{t}{D_o} = \frac{4(6700)}{3(42.85)} = 208 \text{ psig}$$

For values of constant A falling to the left of applicable material/temperature line, the maximum external pressure can be directly evaluated from the following equation:

$$P_a = \frac{2}{3} AE \frac{t}{D_o} \qquad (9.19)$$

where E = modulus of elasticity of material at the temperature being considered.

The external design pressure is 115 psig (the core in full vacuum and external pressure is 100 psig, the pressure acting in the same direction becomes additive).

Since P_a is greater than the design pressure, 115 psig, the assumed thickness is thus satisfactory.

Example 2: The pipe is 4 in., operating under internal pressure of 125 psig at 650°F. The external pressure is 100 psig. The maximum length of spool piece without stiffening rings is 10 ft 0 in. Pipe material is stainless steel ASTM A-312, Type 316. (See Fig. 9.12.)

Calculate the core pipe thickness. Steps 1 to 4 are the same as in Example 1, the results are derived from Figures UGO-28.0 (Fig. 9.8b) and UHA-28.2 (Fig. 9.9a) of Appendix V, ASME Section VIII, Division 1.

From Step 4 in Example 1, the maximum allowable external pressure (P_a) for the assumed sch 10 S core pipe is 208 psig.

External design pressure (P) = 100 psig. Since P_a is greater than P, the assumed thickness is satisfactory.

The external pressure is the governing factor for safe design of a core pipe in the jacketed piping system. If there had not been any external pressure, the pipe carrying 125 psig fluid pressure at 650°F could have thinner wall section.

For calculation of jacket thickness, use Eq. 2.1 (from ASME B31.3, 304.1.2).

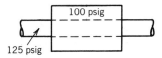

FIGURE 9.12 Jacketed piping (higher pressure in process piping).

TABLE 9.9a SI Metric Conversion Tables: List of SI Units for Use with ASME Boiler and Pressure Vessel Code*

Quantity	Unit	Symbol	Other Units or Limitations
Space and Time			
Plane angle	degree (decimalized)	°	radian
Length	metre	m	
Area	square metre	m²	
Volume	cubic metre	m³	litre (1) for liquid only (use without prefix)
Time	second	s	minute (min), hour (h), day (d), week, and year
Periodic and Related Phenomena			
Frequency	hertz	Hz	
Rotational speed	rev. per second	s⁻¹(rps)	rev. per minute (rpm)
Fluence	nvt		
Neutron energy	Mev	E_n	
Sound (pressure level)	decibel	db	
Mechanics			
Mass	kilogram	kg	
Density		kg/m³	
Moment of inertia		kg·m³	
Force	newton	N	
Moment of force (torque)	newton-metre	N·m	
Pressure and stress	pascal	Pa	(pascal = newton per square metre)
Energy, work	joule	J	kilowatt-hour (kW·h)
Power	watt	W	

162

Impact strength	joule	J
Section modulus	cubic metre	m^3
Moment of section (Second moment of area)		m^4
Fracture toughness	$Pa \cdot \sqrt{m}$	K_{1C}

Heat

Temperature—thermo.†	kelvin	K	degree Celsius (C)
Temperature—other than thermodynamic	degree	C	kelvin (K)
Lin. expansion coeff.		K^{-1}	C^{-1}
Quantity of heat	joule	J	
Heat flow rate	watt	W	
Thermal conductivity		$W/(m \cdot K)$	$W/(m \cdot C)$
Thermal diffusivity	m^2/s		
Specific heat capacity		$j/(kg \cdot K)$	$J/(kg \cdot C)$

Electricity and Magnetism

Electric current	ampere	A
Electric potential	volt	V
Current density		A/m^2
Electrical energy	watt	W
Magnetization current	ampere/metre	A/m

Light

Illumination	lux	lx
Wavelength	angstrom	Å

*Conversion factors between SI units and U.S. customary are given in "ASME Orientation and Guide for Use of Metric Units," and "ASTM E-380."
†Preferred use for temperature and temperature interval is degrees Celsius (C), except for thermodynamic and cryogenic work where kelvins may be more suitable. For temperature interval, 1 K = 1 C exactly.

163

TABLE 9.9b   Commonly Used Conversion Factors (for others see ASTM E-380)

Quantity	To Convert from	To	Multiply by
Plane angle	degree	rd	1.745329 E − 02[†]
Length	in.	m	2.54* E − 02
	ft	m	3.048* E − 01
	yd	m	9.144* E − 01
Area	in.^{2}	m^2	6.4516* E − 04
	ft^2	m^2	9.29034* E − 02
	yd^2	m^2	8.361274 E − 01
Volume	in.^{3}	m^3	1.638706 E − 05
	ft^3	m^3	2.831685 E − 02
	U.S. gallon	m^3	3.785412 E − 05
	Imperial gallon	m^3	4.546090 E − 03
	litre	m^3	1.0* E − 03
Mass	lb (avoirdupois)	kg	4.535924 E − 01
	ton (metric)	kg	1.00000* E + 03
	tons (short 2000 lbm)	kg	9.071847 E + 02
Force	kgf	N	9.80665* E + 00
	lbf	N	4.448222 E + 00
Bending, torque	kgf·m	N·m	9.80665* E + 00
	lbf·in.	N·m	1.129848 E − 01
	lbf·in.	N·m	1.129848 E − 01
	lbf·ft	N·m	1.355818 E + 00
Pressure, stress	kgf/m^2	Pa	9.80665* E + 00
	lbf/ft^2	Pa	4.788026 E − 01
	lbf/in.^{2} (psi)	Pa	6.894757 E + 03
	kips/in.^{2}	Pa	6.894757 E + 06
	bar	Pa	1.0* E − 05
Energy, work	Btu (IT)[‡]	J	1.055056 E + 03
	ft·lbf	J	1.355818 E + 00
Power	hp (550 ft·lbf/s)	W	7.456999 E + 02
Temperature	C	K	$t_K = t_C + 273.15$
	F	K	$t_K = (t_F + 459.67)/1.8$
	F	C	$t_C = (t_F - 32)/1.8$
Temperature interval	C	K	1.0* E + 00
	F	K or C	5.555555 E − 01

*Relationships that are exact in terms of the base units are followed by a single asterisk.

†The factors are written as a number greater than one and less than ten with six or less decimal places. This number is followed by the letter E (for exponent), a plus or minus symbol, and two digits which indicate the power of 10 by which the number must be multiplied to obtain the correct value. For example,

$$3.523907 \, E - 2 \text{ is } 3.523907 \times 10^{-2} = 0.03523907$$

‡International Table.

164

Straight Pipe Under Internal Pressure

The internal pressure design thickness t shall be not less than that calculated by Eq. 2.1, when t is less than $D/6$ (see Chapter 2):

$$t = \frac{PD}{2(SE + PY)} \qquad (2.1)$$

METRIC UNITS

Metric units are used in most countries of the world and piping design codes and standards start dealing with metric units. In a future edition equations given in this book will perhaps be modified to accommodate metric units. All equations given here may be used for piping design in metric units by using proper conversion factors. (See Table 9.9, which is reproduced from Pressure Vessel Code, ASME Section VIII, Division 1). After calculating pipe thickness, diameter, and so on, using British units, the next standard values on the conservative side in metric units may be selected. Caution must be exercised to refer to piping codes and standards in force in each country.

MATERIAL BEHAVIOR AT ELEVATED TEMPERATURE (reference 10)

Elevated temperatures are those at which creep effects are significant. Figure 9.13 is the result of a uniaxial tensile specimen subjected to a load-induced stress level at a given test (low) temperature. Both stress and strain achieve their maximum value at the same time and remain constant at the maximum values thereafter (as long as the load is maintained). When the test temperature is high enough, the strain will increase with time, and possibly until fracture after load application as shown in Figure 9.14. In this case creep effects are significant. If, on the other hand, the elevated temperature uniaxial test is one at which the specimen is initially strained a fixed amount

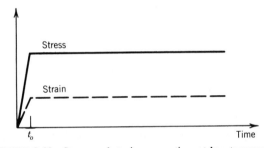

FIGURE 9.13 Stress and strain versus time at low temperature.

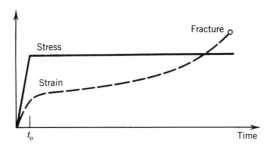

FIGURE 9.14 Stress and strain versus time at elevated temperature (creep effects).

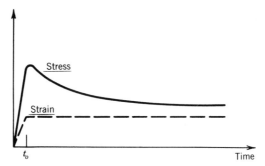

FIGURE 9.15 Stress and strain versus time at elevated temperature (stress relaxation).

and then held constant. The stress–strain history would be somewhat similar to Figure 9.15. The reduction of stress as shown in this figure is called the stress relaxation due to creep effects.

From these figures it can readily be seen that elevated temperature material behavior is a function of stress, temperature, and time.

Application of Creep Data to Piping Design (reference 2)

At temperatures below the creep range, allowable stress values are established at the lowest value of stress obtained from using 25% of the specified minimum ultimate strength at room temperature or 25% of the minimum expected ultimate strength at temperature, or $62\frac{1}{2}$% of the minimum expected yield strength for 0.2% offset at temperature. For bolting material, the stress values were based on 20% of the minimum tensile strength, or 25% of the yield strength for 0.2% offset, whichever is lower. (It is recognized that bolts are always expected to function at stresses above the design value as distinguished from other parts.)

No credit is allowed for any improvement in tensile properties by special heat treatment.

At higher temperatures, where creep governs, the stress values were based on 100% of the stress to produce a creep rate of 0.01% for 1 000 hour, the values so chosen being based on a conservative average of many reported tests as evaluated by the Subcommittee, greater weight being given to longer-time tests in evaluating data. In addition to the above-stated creep-strength requirement, stress values were also limited to 100%[1] of the stress to produce rupture at the end of 100,000 hour, the values so chosen being based on a conservative average as evaluated by the Subcommittee. However, in most cases, the creep strength is far below the rupture strength. Also, in a few cases, the Subcommittee has provided stress values without any rupture test data on the specific composition, such approval being based on tests of materials of similar composition.

In the transition range of temperatures, the stress allowances were limited to values obtained from a smooth curve joining the values for the low- and high-temperature ranges, the curve lying on or below the curve of $62\frac{1}{2}\%$ of the minimum expected yield strength at temperature.

In the choice of stress values in the range where a percentage of the tensile strength or yield strength governs, the limitations indicated above have been waived in certain cases, identified by a footnote,[1] as it was felt that higher stress values might be justified when deformation was not in itself objectionable, provided all other requirements were met.

In the design of equipment not covered by codes, the design stress values may be decided upon by the manufacturer and purchaser of the piping, and should be based on the best available data plus a knowledge of the expected life of the equipment as well as the operating conditions and the possible hazard to personnel. Rules generally followed are:

a. Up to 750 or 850°F, 25% of the short-time tensile strength and not exceeding $62\frac{1}{2}\%$ yield strength.

b. Above 900°F, 100% of the stress to produce a second-stage creep rate of 0.01% in 1000 h, or 80% of the stress to produce rupture in 100,000 h, whichever is lower.

REFRACTORY LININGS

Refractory linings are used in kilns, coke ovens, furnaces, and stacks to protect metal parts from direct exposure of very high temperatures (Fig. 9.16). Refractories need to withstand very high temperatures without melting, should have necessary mechanical and heat transfer properties, should not react with the medium inside the furnace, and large quantities need to be available at low prices (reference 5). Based on the chemical property, refractories are classified as acidic (example, silica), basic (magnesia), and

---

[1]This 100% value pertains to the Unfired Pressure Vessel Code. In the Power Boiler Code this stress limitation is 60% of the average or 80% of the minimum stress to produce rupture in 100,000 h as reported by test data.

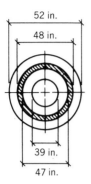

52 in.

48 in.

39 in.

47 in.

FIGURE 9.16   Refractory lining.

neutral (chrome ore). The standard form for refractories is brick. In steel mill design, a large diameter, inclined pipes with refractory lining are used. Calculation of weight to be supported at support points is of importance (Fig. 9.16).

Density of refractory	$= 40 \, \text{lb/ft}^3$
Density of steel	$= 0.283 \, \text{lb/in.}^3$
Density of insulation	$= 11 \, \text{lb/ft}^3$
Inside diameter of refractory	$= 39$ in.
Outside diameter of pipe	$= 48$ in.
Outside diameter of insulation	$= 52$ in.

Weight per foot = weight of (refractory + pipe metal + insulation)   (9.20)

$$= \frac{\pi}{4}(47^2 - 39^2)12\left(\frac{40}{1728}\right) + \frac{\pi}{4}(48^2 - 47^2)12(0.283)$$

$$+ \frac{\pi}{4}(52^2 - 48^2)12\left(\frac{11}{1728}\right)$$

$$= 151.55 + 255.83 + 21$$

$$= 428.46 \, \text{lb/ft}$$

EXERCISES

1. True or false?
   (a) Cold springs cannot be used to reduce the moment on a vessel.
   (b) Hot modulus of elasticity can be used for stress calculation.
   (c) For span calculation, maximum deflection allowed inside plant is 1 in.

(d)   The piping code B31.1 allows use of inplane and outplane SIF.

(e)   Pad is needed when area required is smaller than area removed.

(f)   Internal pressure when included decreases the SIF value for an elbow.

(g)   When the effects of flanges are included for an elbow, the flexibility factor reduces.

(h)   Expansion loops take less space compared with expansion joints.

2.   True or false?

(a)   Cold spring can be applied only to hot piping.

(b)   For span calculation hot modulus should be used.

(c)   Internal pressure increases flexibility factor value for elbow.

(d)   Expansion loops are safer than expansion joints.

(e)   The vessel nozzle growth is to be included in stress calculations.

(f)   Piping (refinery) design is governed by ASME code.

(g)   Outside diameter of 6 in. nominal pipe is $6\frac{1}{2}$ in.

(h)   The exit diameter is larger than the inlet diameter in pressure relief valves.

3.   An underground pipeline with ASTM-A53 Grade B material and 12 in. sch 40 pipe has the following conditions:

   Operating temperature: 175°F

   Installed temperature: 70°F

   Depth of burial: 3 ft 6 in.

   Specific gravity of content: 0.73

   Operating pressure: 375 psig

   (a)   Where is the location of the natural anchor?   (b) The amount of thermal expansion at the end? (The friction coefficient = 0.75.)

4.   Calculate dynamic reaction force with open discharge system.

   Flow = 165,000 lb/hr

   Area of valve orifice = 55 sq in.

   $J = 778$ ft-lb/Btu; $g = 32.2$ ft/sec^2

   Temperature = 650°F; $PR = 155$ psig

   $h_o = 1374$ Btu/lb (from steam tables)

   $a = 823$ Btu/lb; $b = 4.33$

   Discharge line is 6 in. sch 40

5.   A double-acting reciprocating gas compressor has a maximum rated speed at 650 rpm with a pulsation pressure limited to 13 psi. The discharge pipe is 3 in. sch 40. What will be the pulsation force? What is the maximum span of pipe support? $I_p = 3.02$ in.^{4}

6.   A 20 in. carbon steel pipe has a wall thickness of 0.25 in. The minimum specified yield stress is 47,000 psi at a design pressure 600 psig. The design temperature is 170°F and the winter temperature is 25°F. If the

bending stress is 9200 psi, what is the tie-in temperature? (The temperature range between summer and winter is 25 to 100°F.)

7. A crude pipeline of 18 in. diameter is to be designed with an operating pressure of 1300 psi and an operating temperature of 170°F. It is decided that API-5Lx, Grade X52 electric-resistance-welded pipe will be used. The joint efficiency of the weld is 85%. The specified minimum yield stress is 53,000 psi. Construction temperature is expected to be 75°F. If the bending stress is 9750 psi, what will be the pipe wall thickness?

8. If the pipe of Exercise 7 will be fully restrained, what will be the longitudinal stress at anchor point?

9. For a 20 in. pipeline the required maximum operating pressure is 670 psig and the maximum expected operating temperature is 165°F. The material of pipe will be API-5Lx, Grade X52 with a specified minimum yield stress of 49,000 psi. Based on pressure the calculated wall thickness is 0.25 in. If the tie-in temperature has to be at 75°F with a bending stress 7700 psi, calculate stress in the pipe.

REFERENCES

1. Information on valves are available from:
    Vogt Forged Steel Valves, Fittings, Union, Flanges, Catalog F-12.
    Powell Valves, Catalog 75
    Walworth Valves, Catalog 130
    Crane Valve Fittings, Catalog No. 60
2. King, C. Reno and Sabin Croker. *Piping Hand Book*, New York: McGraw-Hill.
3. Amir, S. J. "Calculating Heat Transfer from a Buried Pipeline," *Chemical Engineering* (August 4, 1975).
4. Schweitzer. *Handbook of Corrosion Resistant Piping*, Industrial Press.
5. Nord, Melvin. *Textbook of Engineering Materials*, New York: Wiley.
6. ANSI Standard A58.1, *Wind Loads for Buildings and Other Structures*.
7. Leonards, G. A. *Foundation Engineering*, New York: McGraw-Hill.
8. Owens-Corning, *Plastic Pipe Program*.
9. ITT Grinnell. "Pipe Hanger Design and Engineering," in *Weight of Piping Materials*, revised in 1979.
10. Truong, Q. N. "International Piping Conference", at Texas A & M University, Texas, April, 1983.

CHAPTER TEN

NUCLEAR COMPONENTS CODE ASME SECTION III

DESIGN LOADS AND SERVICE LIMITS

Nuclear Components Design Code, ASME Section III (reference 5 in Chapter 4) requires effects of earthquake to be included in the design of piping, piping supports, and restraints (see Section III, subsection NC-3622, Dynamic Effects). The loadings, the movements including earthquake anchor movements, and number of cycles to be used in the analysis are part of the design specifications. The stresses resulting from these earthquake effects must be included with weight, pressure or other applied loads when making the required analysis. Section III also requires design loadings (NC-3611.2(b)), and service loadings (NC-3611.2(c)) be specified. Service loadings are grouped as Level A (Eqs. 10.10 and 10.11), Level B (Eq. 10.9U), Level C (Eq. 10.9E), and Level D (Eq. 10.9F) service limits. See also Chapter 6 for a brief explanation of these service limits. Service limits B, C, and D require inclusion of earthquake loads. Design loading is given by Eq. 10.8. Equation numbers 10.4 to 10.7 have been eliminated so that equation numbers used in (Table 10.2) will be the same as those used in Section III (reference 5 in Chapter 4)

Nuclear piping is classified as Class 1 (NB), Class 2 (NC), and Class 3 (ND), piping. The piping connecting the reactor and the steam generator and other critical piping comes under Class 1 analysis, which is beyond the present scope of this book. Sample analysis of a Class 1 Nuclear Piping System prepared by ASME Boiler and Vessel Committee would be helpful for further reading. Design of Class 2 is presented here. Most companies conservatively design Class 3 piping under Class 2 guidelines.

FLEXIBILITY AND STRESS INTENSIFICATION FACTORS

Table 10.1 gives equations for calculating flexibility characteristics h, flexibility factor k, and stress intensification factor i for Class 2 piping. This

171

TABLE 10.1 Stress Intensification Factor (NC-3000 Design)

Description	Flexibility Characteristic h	Flexibility Factor k	Stress Intensification Factor i	Sketch (Fig. NC-3673.2(b)-1)
Welding elbow or pipe bend (1) (2) (3)[a]	$\dfrac{t_n R}{r^2}$	$\dfrac{1.65}{h}$	$\dfrac{0.9}{h^{2/3}}$	
Closely spaced miter bend (1) (2) (3) $s < r\,(1 + \tan\theta)$	$\dfrac{s t_n \cot\theta}{2r^2}$	$\dfrac{1.52}{h^{5/6}}$	$\dfrac{0.9}{h^{2/3}}$	
Single miter or widely spaced miter bend (1) (2) (4) $s \geq r(1 + \tan\theta)$	$\dfrac{t_n(1 + \cot\theta)}{2r}$	$\dfrac{1.52}{h^{5/6}}$	$\dfrac{0.9}{h^{2/3}}$	

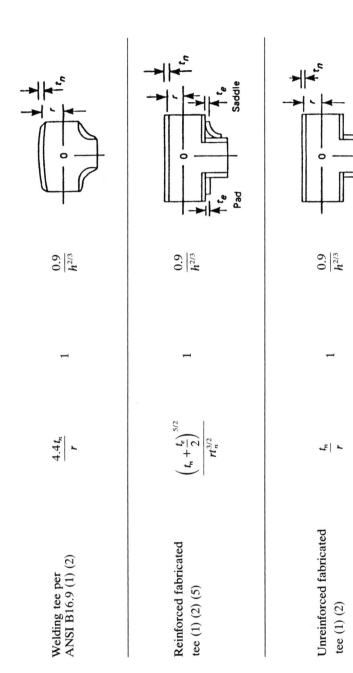

	h	Flexibility factor k	Stress intensification factor i
Welding tee per ANSI B16.9 (1) (2)	$\dfrac{4.4\, t_n}{r}$	1	$\dfrac{0.9}{h^{2/3}}$
Reinforced fabricated tee (1) (2) (5)	$\dfrac{\left(t_n + \dfrac{t_e}{2}\right)^{5/2}}{r\, t_n^{3/2}}$	1	$\dfrac{0.9}{h^{2/3}}$
Unreinforced fabricated tee (1) (2)	$\dfrac{t_n}{r}$	1	$\dfrac{0.9}{h^{2/3}}$

ᵃSee notes following tables.

173

TABLE 10.1 *Continued*

Description	Flexibility Factor k	Stress Intensification Factor i	Sketch
Branch connection (6)	1	$1.5\left(\dfrac{R_m}{T_r}\right)^{2/3}\left(\dfrac{r'_m}{R_m}\right)^{1/2}\left(\dfrac{T_b}{T_r}\right)\left(\dfrac{r'_m}{r_p}\right)$	Figure 10.1 Fig. NC-3673.2(b)-2, Sec III
Butt weld (1) $t_n \geq \frac{3}{16}$ and $\dfrac{\delta}{t_n} \leq 0.1$	1	1.0	
Butt weld (1) $t_n \leq \frac{3}{16}$ or $\dfrac{\delta}{t_n} > 0.1$	1	1.0 for flush weld 1.8 for as-welded	
Fillet-welded joint, socket-welded flange, or single-welded slip on flange	1	2.1	Fig. NC-3673.2(b)-3, sketches (a), (b), (c), (e) and (f)
Brazed joint	1	2.1	Fig. NC-4511-1
Full fillet weld	1	1.3	Fig. NC-3673.2(b)-3, sketch (d)

174

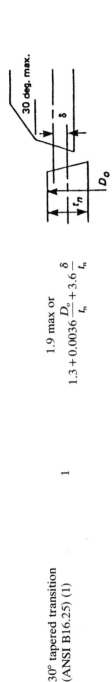

30° tapered transition (ANSI B16.25) (1)	1	1.9 max or $$1.3 + 0.0036\,\dfrac{D_o}{t_n} + 3.6\,\dfrac{\delta}{t_n}$$
Concentric reducer (ANSI B16.9 or MSS SP48) (7)	1	2.0 max or $$0.5 + 0.01\alpha\left(\dfrac{D_2}{t_2}\right)^{1/2}$$
Threaded pipe joint or threaded flange	1	2.3
Corrugated straight pipe or corrugated or creased bend (8)	5	2.5

1. The following nomenclature applies:

r = mean radius of pipe, in. (matching pipe for tees and elbows)
t_n = nominal wall thickness of pipe, in. (matching pipe for tees and elbows, see note (9))
R = bend radius of elbow or pipe bend, in.
θ = one-half angle between adjacent miter axes
s = miter spacing at centerline, in.
t_e = reinforced thickness, in.
δ = mismatch, in.
D_o = outside diameter, in.

175

TABLE 10.1 *Continued*

2. The flexibility factors k and stress intensification factors i apply to bending in any plane for fittings and shall in no case be taken less than unity. Both factors apply over the effective arc length (shown by heavy centerlines in the sketches) for curved and miter elbows, and to the intersection point for tees. The values of k and i can be read directly (Table 4.3b) by entering with the characteristic h computed from the equations given.

3. Where flanges are attached to one or both ends, the values of k and i shall be *corrected* by the factor c given below, which can be read directly from (Table 4.3b), entering with the computed h.

$$\text{One end flanged } c = h^{1/6} \qquad \text{Both ends flanged } c = h^{1/3}$$

4. Also includes single miter joints.

5. When $t_e > 1.5 t_n$, $h = 4.05 t_n / r$.

6. The equation applies only if the following conditions are met:
   (a) The reinforcement area requirements of NC-3643 are met.
   (b) The axis of the branch pipe is normal to the surface of run pipe wall.
   (c) For branch connections in a pipe, the arc distance measured between the centers of adjacent branches along the surface of the run pipe is not less than three times the sum of their inside radii in the longitudinal direction or is not less than two times the sum of their radii along the circumference of the run pipe.
   (d) The inside corner radius r_1 (Fig. NC-3673.2(b)-2) (Fig. 10.1) is between 10 and 50% of T_r.
   (e) The outer radius r_2 is not less than the larger of $T_b / 2$, $(T_b + y)/2$ (Fig. NC-3673.2(b)-2 (Fig. 10.1) sketch (c)) or $T_r / 2$.
   (f) The outer radius r_3 is not less than the larger of
       (1) $0.002 \theta d_o$
       (2) $2 (\sin \theta)^3$ times the offset for the configurations shown in Figs. NC-3673.2(b)-2 (Fig. 10.1) sketches (a) and (b).
   (g) $R_m / T_r \leq 50$ and $r_m / R_m \leq 0.5$.

7. The equation applies only if the following conditions are met:
   (a) Cone angle α does not exceed 60 deg., and the reducer is concentric.
   (b) The larger of D_1 / t_1 and D_2 / t_2 does not exceed 100.
   (c) The wall thickness is not less than t_1 throughout the body of the reducer, except in and immediately adjacent to the cylindrical portion on the small end, where the thickness shall not be less than t_2.

8. Factors shown apply to bending; flexibility factor for torsion equals 0.9.

9. The designer is cautioned that cast butt welding elbows may have considerably heavier walls than that of the pipe with which they are used. Large errors may be introduced unless the effect of these greater thicknesses is considered.

176

table is reprinted from ASME Section III, Table NC-3000. Note that only one equation for SIF corresponding to the higher value of $0.9/h^{2/3}$ is given. Nuclear Code also gives equations to calculate SIF for branch connection based also on the branch dimensions, welded joints, 30° taper transition, and for concentric reducer.

Examples

1. Calculate SIF for a given weld boss (same equations apply for socket weld half coupling and weld couplet) branch connection. Run or header is 12 in. diameter with thickness of 0.375 in. Branch is 6 in. diameter with wall thickness of 0.28 in. Outside diameter of weld boss = OD of branch + 2 (thickness of weld boss).
    The SIF A is given by:

$$\text{SIF} = 1.5 \left(\frac{R_m}{T_r}\right)^{2/3} \left(\frac{r'_m}{R_m}\right)^{1/2} \left(\frac{T_b}{T_r}\right)\left(\frac{r'_m}{r_p}\right)$$

$$\text{SIF} \geq 1.0 \qquad \frac{R_m}{T_r} \leq 50 \qquad \frac{r'_m}{R_m} \leq 0.5. \tag{10.1}$$

where R_m = mean radius of run pipe, inches
r'_m = mean radius of branch pipe, inches
T_r = nominal wall thickness of run pipe, inches
T_b = nominal thickness of branch pipe, inches
r_p = outside radius of coupling or boss, inches
    (outside radius of branch connection)
Figure 10.1 (from ASME Section III, subsection Figure NC-3673.2(b)2) shows four branch connections for which Eq. 10.1 applies.
    Run OD = 12.75 in. and thickness = 0.375 in.
    Branch OD = 6.625 in. and thickness = 0.28 in.

$$R_m = \frac{12.75 - 2(0.375)}{2} = 6.025 \text{ in.}$$

$$r'_m = \frac{6.625 - 2(0.28)}{2} = 3.0325 \text{ in.}$$

$$r_p = \text{outside radius of weld boss} = [6.625 + 2(\tfrac{5}{8})]\tfrac{1}{2} = 3.9375 \text{ in.}$$

$$\text{SIF}_A = \left(\frac{6.025}{0.375}\right)^{2/3} \left(\frac{3.0325}{6.025}\right)^{1/2} \left(\frac{0.280}{0.375}\right)\left(\frac{3.0325}{3.9375}\right) 1.5$$

$$= (16.066)^{0.666}(0.5033)^{0.5}(0.7466)(0.770)1.5$$

$$= (6.367)(0.709)(0.7466)(0.770)1.5$$

$$= 3.89$$

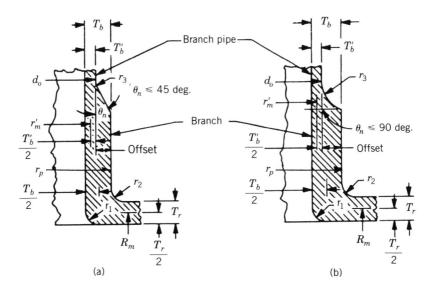

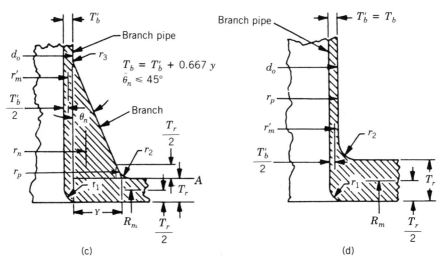

FIGURE 10.1   Branch dimensions (ASME Section III, NC 3673.2(b)-2).

The above Equations apply only if conditions of Note 6 with Table 10.1 are met. Note that $r'_m/R_m = 0.5033$, which exceeds the limit of 0.5 slightly.

The stress intensification factor should be taken as the higher of the value calculated above by the equation and the SIF for the branch pipe. The SIF for the straight pipe assuming the socket weld will be equal to 2.3.

2. Calculate SIF for concentric reducer with a larger diameter of 6.625 in. and smaller diameter of 4.5 in. Thickness on the larger side is 0.280 in. and smaller side is 0.237 in. (Fig. 10.2).

$$\text{Cone angle of reducer} = \alpha = \sin^{-1}\left(\frac{D_1 - D_2}{2L}\right)$$

$$= \sin^{-1}\left(\frac{6.625 - 4.5}{2(5.5)}\right)$$

$$= \sin^{-1}(0.19318)$$

$$\alpha = 11.138 \text{ deg}$$

Check for use of the SIF equation in Table 10.1, Note 7:

$$\frac{D_1}{t_1} = \frac{6.625}{0.280} = 23.66 < 100 \quad \text{OK}$$

$$\frac{D_2}{t_2} = \frac{4.5}{0.237} = 18.98 < 100 \quad \text{OK}$$

$$\alpha = 11.138° < 60° \quad \text{OK}$$

$$\text{SIF} = 0.5 + 0.01\alpha\sqrt{\left(\frac{D_2}{t_2}\right)} \quad \text{or} \quad \leq 2.0 \qquad (10.2)$$

$$= 0.5 + 0.01(11.138)\sqrt{\left(\frac{4.5}{0.237}\right)}$$

$$= 1.00$$

Use

$$\text{SIF} = 1.00$$

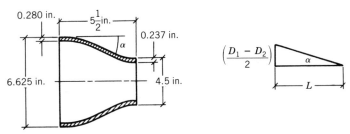

FIGURE 10.2   Concentric reducer.

TABLE 10.2 Criteria for Class 2 Piping Stress Evaluation[4] (category I and IL piping)[11]

NC-3652 Eq. No.	Plant Condition and Load Source[1]	Moment from[1]	Code Stress[10]	Allowable Stress	Text Eqs.
8	Design and normal[7] Pr + sustained	$M_A = M$ (DW, CS, PL, PT)[2]	$\dfrac{Pd^2}{D_o^2 - d^2} + \dfrac{0.75iM_A}{Z}$	S_h	(10.8)
9U[3]	Upset[7] Pr + sustained + occasional	$M_A = M$ (DW, CS, PL) $M_{BU} = M$ (E1, VT, WH)	$\dfrac{P_{\max}d^2}{D_o^2 - d^2} + \dfrac{0.75i}{Z}(M_A + M_{BU})$	$1.2S_h$	(10.9U)
9E	Emergency Pr + sustained + occasional	$M_A = M$ (DW, CS, PL) $M_{BE} = M$ (E1, VT, WH)	$\dfrac{P_{\max}d^2}{D_o^2 - d^2} + \dfrac{0.75i}{Z}(M_A + M_{BE})$	$1.8S_h$	(10.9E)
9F[3]	Faulted Pr + sustained + occasional	$M_A = M$ (DW, CS, PL) $M_{BF} = M$ (E2, VT, WH, DM, JI)[5]	$\dfrac{P_{\max}d^2}{D_o^2 - d^2} + \dfrac{0.75i}{Z}(M_A + M_{BF})$	$2.4S_h$	(10.9F)
10	Normal or upset (secondary) Thermal expansion + occasional or	$M_C = M$ (Ti, S1[6], or S2[6])	$\dfrac{iM_C}{Z}$	S_A	(10.10)
11	Pr + sustained + expansion	$M_A = M$ (DW, CS, PL) $M_C = M$ (Ti, S1, or S2)	$\dfrac{Pd^2}{D_o^2 - d^2} + \dfrac{0.75iM_A}{Z} + \dfrac{iM_C}{Z}$	$(S_A + S_h)$	(10.11)
and					
10a[8]	Anchor movement	$M_D = M$ (BS)[9]	$\dfrac{iM_D}{Z}$	$3S_C$	(10.10a)

PR 9+10	Pipe rupture	$\dfrac{P_{max}d^2}{D_o^2 - d^2} + \dfrac{0.75i}{Z}(M_A + M_{BU}) + \dfrac{iM_C}{Z}$	$0.8(1.2S_h + S_A)$	(10.12)
AV 9+10	Active valve	$\dfrac{P_{max}d^2}{D_o^2 - d^2} + \dfrac{1}{Z}(M_A + M_B) + \dfrac{1}{Z}M_C$	(AVC) (S_y)	(10.13)

Pipe rupture: $M_A = M$ (DW, CS, PL), $M_{BU} = M$ (E1, VT, WH)

Active valve: $M_B = \max(MBU, MBE, MBF)$, $M_A = M$ (DW, CS, PL)

Notes

1. All moments are calculated as three components M_x, M_y, M_z. The terms M_A, M_B, M_C represent the square root of the sum of the squares (SRSS):

$$M_A, M_B, M_C = \sqrt{M_x^2 + M_y^2 + M_z^2}$$

2. Cold spring should not be used to reduce stress. Cold spring loads may be considered in load evaluation on support and equipment. Preload may be applied at spring supports to relieve nozzle load. Pressure thrust (in case of expansion joint bellows without tie rods) is included with the dead weight loads.

3. Evaluation of Eqs. 10.9U and 10.9F is not required for category IL piping (where only limited structural integrity is required).

4. The ASME Code Section III, Division 1, subsection NC-3652, Analysis of Piping.

5. In general, the jet impingement loads are not available at the analysis phase and thus not included in the evaluation for faulted condition.

6. The term SAM also known as dynamic anchor movements S1 and S2 may be included in Eq. 10.9 or 10.10 but *not* both.

7. For normal or upset condition, Eq. 10.8 or 10.9U plus Eq. 10.10 or 10.11 must be satisfied.

8. Equation 10.10a applies to nonrepeated movement such as building settlement.

9. Moment M_D for emergency or faulted (secondary) load case in Eq. 10.10a must include both containment thermal movement (CT) and containment pressure movement (CP) after a design basis accident (DBA).

10. Pressure stress can also be calculated by using formula $PD_o/2$(thickness).

11. Table 10.2 can be used conservatively for Class 3 piping (ND components) also.

181

ANALYSIS FOR CLASS 2 (NC COMPONENTS) PIPING STRESS EVALUATION

Table 10.2 gives criteria for rigorous or comprehensive analysis for Class 2 piping (Reference 1). The explanation of the abbreviations used in the table is as follows:

$\quad$ DW = deal weight

$\quad\,$ CS = cold spring

$\quad\,$ PL = preload

$\quad\,$ VT = valve thrust

$\quad$ WH = water hammer

$\quad\,$ E1 = operating basis earthquake load (OBE)

$\quad\,$ E2 = safety shutdown earthquake load (SSE)

$\quad\,$ JI = jet impingement

$\quad\,$ S1 = seismic anchor movement due to OBE

$\quad\,$ S2 = seismic anchor movement due to SSE

$\quad$ BS = building settlement

$\quad\,$ Ti = thermal load

$\quad$ CP = containment movement due to pressure after DBA

$\quad$ CT = containment movement due to temperature after DBA

$\quad\quad P$ = design pressure, psi

$\quad\, D_o$ = outside diameter of pipe, inches

$\quad\;\, d_i$ = inside diameter of pipe, inches

$\quad\quad\, i$ = stress intensification factor

$\qquad\;\;$ ($i \geq 1$, $0.75i \geq 1.0$)

$\qquad\;\;$ 0.75 i cannot be less than 1.0

$\quad\; Z$ = modulus of section, in.^{3}

AVC = active valve coefficient (0.75 to 0.9)

$\,$ DM = dynamic movement

DBA = design basis accident

$\quad\; S_c$ = basic allowable stress at minimum (cold) temperature, psi

$\quad\, S_h$ = basic allowable stress at design temperature, psi (see Sec. III for values)

$\quad\; S_A$ = allowable stress range (Eq. 4.1)

$P_{\max}$ = peak pressure, psi

$\quad\; S_y$ = yield stress, psi

$\quad$ PT = bellows pressure thrust, lb

M_{BF} = M_{B} under faulted condition, ft. lb.

NATURAL FREQUENCY

If the natural frequency of a piping system is at or near the frequency of an exciting source, for instance, a compressor, the resulting amplitudes may induce bending stresses that lead to premature fatigue failure. A necessary design criterion must be therefore that the natural frequencies in a piping system must be significantly higher than or different from the frequencies of the exciting source.

Natural frequency in cycles per second is given by Eq. 10.3.

$$f_n = \frac{\alpha}{L^2} \sqrt{\frac{EI}{W}} \qquad (10.3)$$

where L = length of pipe, feet

E = modulus of elasticity, psi

I = moment of inertia, in.^{4}

W = weight of pipe, lb/ft

α = value depending upon end conditions and the mode under consideration. See Table 10.3 for values of α.

TABLE 10.3   Natural Frequency Calculation (α Value for Eq. 10.3)

End Condition	Configuration	Mode	Value of α
Both ends simply supported		Fundamental(1st)	0.743
		Second mode	2.97
Both ends fixed		First mode	1.69
		Second mode	4.64
One end fixed; one end simply supported		First mode	1.16
		Second mode	3.76

PIPING SYSTEMS TO BE ANALYZED

Stress analysis will normally be performed for piping systems in the following categories:

1.   Lines 3 in. and larger (a) connected to rotating equipment, or (b) subject to differential settlement of connected equipment and/or supports, or (c) with temperatures less than 20°F

2. Lines connected to reciprocating equipment
3. Lines 4 in. and larger connected to air coolers, steam generators, or fired heater tube sections
4. Lines 6 in. and larger with temperatures of 250°F and higher
5. Lines with temperatures of 600°F and higher
6. Lines 16 in. and larger
7. Alloy lines
8. High pressure lines
9. Lines subject to external pressure
10. Thin-walled pipe or duct of 18 in. diameter and over, having an outside diameter over wall-thickness ratio of more than 90
11. Lines requiring proprietary expansion devices, such as expansion joints and victaulic couplings
12. Underground process lines
13. Internally lined process piping
14. Lines in critical service
15. Pressure relief systems

Information Needed for Pipe Stress Analysis

1. Outside diameter of piping, wall thickness (or nominal diameter, sch number) (Appendix A4)
2. Temperature, internal pressure
3. Material of piping. (Expansion coefficient, Young's modulus, and material density will be selected for this material.) (Appendix A2)
4. Insulation thickness and insulation material. (If not given, standard thickness for calcium silicate will be selected.)
5. Specific gravity of contents
6. Any wind load to be considered? If yes, the direction of application is important.
7. Any anchor initial translation, Δx in inches, Δy in inches, Δz in inches. (For towers, exchangers, and so on, nozzle initial translation is important.)
8. Corrosion allowance for piping, inches
9. Flange rating, psi (ANSI B16.5)
10. Standard valve weight and flange weight will be (Reference 1 in Chapter 9) used. (For special valves mark the weight on pipe stress isometric.)

11. Long radius elbows will be used. (If short radius or any other bend radius, mark on the isometric.) For short-radius elbow, radius = diameter

12. Any allowable loading from manufacturers on pumps, turbines, compressors? (From the vendor drawing for equipment.)

13. Any preference to use expansion loops, expansion joints, and so on, if needed? (Chapter 5)

14. Mark type of intersection (reinforced fabricated tee, etc.)

15. Mark support locations (available steel crossing, and so on) on the isometric

16. Is hydraulic testing load condition to be considered to get structural support loads? (Eq. 2.7)

17. Pipe stress isometrics (x-, y-, z-axis) piping plans, and sections are necessary.

USEFUL HINTS IN PIPING DESIGN

1. Maximum movements at branch location must be lower than specified limit. The branch line should be laid with enough flexibility to absorb the header movement.

2. In nuclear piping analysis, the branch also needs to be included with the header if the area moment of inertia ratio $I_h/I_b < 40$. In other words decoupling is not allowed. Check company criteria for the ratio to use.

3. If branch pipe is analyzed separately, the movements at the decoupling should be included as initial (or imposed) movement in the branch line calculation.

4. The modulus of elasticity value at operating temperature may be used for piping to calculate the loads at equipment as per standard API 610 (reference 7 in Chapter 7). Using E_{hot} will result in lower loads because E_{hot} is lower than E_{cold}. The piping is more flexible when E value is lower.

5. The guide should not be located close to the change in direction. A minimum leg is required for absorbing the expansion. Calculate the minimum leg as per method outlined in Chapter 1.

6. Credit cannot be obtained for cold spring in stress calculation. Only loads at the equipment may be reduced by including the effect of cold spring.

7. Provide longer pipe support shoes when axial deflections are large.

8. Nuclear Regulatory Commission issues regulatory guides to be followed in design.

Results

The results or commonly known as output from computer-aided analysis generally consist of the following:

1. *From Input*: Coordinates of nodes or data points, length, diameter, thickness, bend radius, total weight of pipe, temperature, expansion coefficient, modulus of elasticity, pressure, valve weights, lengths, wind loads, support location, and types.
2. *Results*: Deflections, rotations, forces, moments, SIF, resultant bending stress, torsional stress, and expansion stress.
3. Requirements for different codes vary. The ASME/ANSI B31.3 compliance is discussed next.
   (a) Wall thickness used should be greater than the minimum thickness required using Eq. 2.1.
   (b) Pressure input should be lower than the allowed pressure calculated using Eq. 2.4.
   (c) Expansion stress S_E calculated using Eq. 4.7 should be less than the expansion stress range from Eq. 4.1 or 4.2.
   (d) Expansion stress does not include either weight or pressure loading but only thermal loads.
   (e) The additive stress S_L should not exceed hot stress S_h. S_L = resultant bending stress from weight loads + longitudinal pressure stress, S_{LP}:
   where

$$S_{LP} = \frac{(\text{pressure})(\text{OD})}{4(\text{thickness})}$$

If the piping system is overstressed or if equipment nozzle loads are excessive, then flexibility of piping system needs to be improved as discussed in Chapters 1 and 5.

4. Actual deflection (maximum considering different load cases) should be lower than sleeve clearance.
5. Stress ratio is the ratio of code stress (Table 10.2) to allowable stress and should be less than 1.0.

COMPUTER MODELING

Different computer programs suggest the inputing (input coding) of different piping components differently. The following description outlines one computer model to code a piping element. However, based on the assumptions of a program logic, coding could be different. Information about the following will assist computer modeling.

1. Initial anchor movements, described later in detail
2. Type of intersections (see Figures 10.6 through 10.11)
   Reducing tee
   Fabricated tee
   Unreinforced tee
   Reinforced tee, pad or saddle
   Weldolet
   Sockolet
   Sweepolet
   Pipet
   Latrolet
   Socket weld tee
3. Elbows, bends, miter bends, elbolet, socket weld elbow, elbolet, support on the bend, and flanged elbows (see Figures 10.17 through 10.18)
4. Concentric and eccentric reducers, reducing insert, and half coupling (see Figures 10.13 through 10.16)
5. Cold spring, cut short or cut long (Eq. 4.12)
6. Wind loading (Reference 6, Chapter 9)
7. Valves, flanges, valve operators, cap, blind flanges (see Figures 10.19 through 10.21)
8. Releasing anchor for a specific direction of the restraint, flexible anchor, spring rate inclusion at nozzle anchors (Eq. 7.3)
9. Expansion joints (single bellows, gimbal, hinged, universal), pressure thrust calculation (Eq. 5.4)
10. One directional supports
11. Insulation weight, content weight, refractory weight (Eq. 9.20)
12. Loop closure, coordinates of balance points
13. Jacketed piping (Figure 9.11)

INITIAL ANCHOR MOVEMENTS AND SUPPORT MOVEMENTS

Anchors and supports are moved by a calculated amount in the analysis to include:

1. Movement due to thermal growth of towers, heat exchangers, pumps, turbines, and so on

2. Building settlement, tank settlement (may occur when piping is cold)
3. Seismic anchor and support movements known as SAM

Movements Due to Thermal Growth

Vertical Vessels

Figure 10.3 shows a vessel with nozzles with different orientations. Calculated thermal movement is based on the mean temperature and length of the vessel section. It is not unusual to have many different temperatures at different elevations.

Thermal coefficient for the temperatures is obtained from Appendix A1, and shown in Table 10.4.

The vessel material is carbon steel and diameter is 72 in.

$$\text{Vertical thermal growth at nozzle A} = 12(0.0099) + 14(0.014)$$
$$+ 8(0.027) + 6(0.0411) + 4(0.0563)$$
$$= 1.002 \text{ in.}$$

$$\text{Growth at nozzle B} = 0.3148 + 3(0.027) = 0.3958 \text{ in.}$$

$$\text{Horizontal radial growth at B} = \left(\frac{72}{2 \times 12}\right)(0.027) = 0.081 \text{ in.}$$

$$\text{Growth at C} = 0 \text{ in.}$$

$$\text{Vertical growth at support D} = 0.53 + 2.5(0.0411) = 0.632 \text{ in.}$$

Flexibility problems are severe when the vessel is hot and the piping is cold. Elevation difference between nozzle A and support D should be minimum to avoid large growth differential and thus avoid a spring support. If the support is built from structural steel (cold), a spring at this support location is necessary. When supported from the vessel, the support design

TABLE 10.4   Thermal Coefficient
(for Fig. 10.3)

Temp. °F	in./100 ft
70	0.00
200	0.99
250	1.40
400	2.70
550	4.11
700	5.63

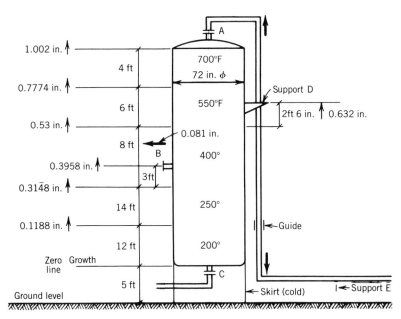

FIGURE 10.3 Thermal growth at the vessel.

load at D is critical and the vessel shell needs checking for local stress. From support location D, the pipe above it grows up and the piping below it grows downward. The first rigid support, shown as E, should not be located close to the drop to absorb the downward growth.

Heat Exchangers

Figure 10.4 shows a heat exchanger. The important thing concerning the exchangers is finding the base support that is anchored (A1) and the other support that is slotted shown as S1 in Figure 10.5. The base with slots is

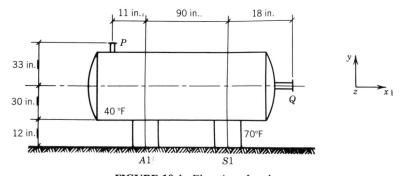

FIGURE 10.4 Elevation of exchanger.

FIGURE 10.5 Anchored and slotted supports for heat exchanger.

allowed to slide along the axis of the exchanger ($+x$ direction in Figure 10.4). The selection of which one of the two is anchored could be based on the growth of the connecting pipe. It is necessary that the exchanger grow with the piping.

In Figure 10.4 the shell temperature is only 40°F and the shell contracts instead of expanding. The coefficient of expansion is 6.07 in./in./°F.

$$\text{Vertical growth at } P = (+63)(6.07 \times 10^{-6})(40-70) = -0.0115 \text{ in.}$$

(The negative sign shows that the shell contracts downward, $\Delta y = -0.0115$ in.)

$$\text{Horizontal growth at } P = \Delta x = (-11)(6.07 \times 10^{-6})(40-70) = 0.002 \text{ in.}$$

(The minus sign is used for 11 in. because P is on the negative x side of anchor $A1$ in which horizontal growth starts.)

$$\Delta x \text{ at } Q = (108)(6.07 \times 10^{-6})(40-70) = -0.0197 \text{ in.}$$
$$\Delta y \text{ at } Q = 30(6.07 \times 10^{-6})(40-70) = -0.0055 \text{ in.}$$

Δx in movement in x direction.

Sometimes two heat exchangers are stacked one over the other or right next to each other. The expansion between the exchangers is critically important.

MODELING OF PIPING ELEMENTS

The modeling of piping elements described for Figures 10.6 through 10.21 is from the Tennessee Valley Authority (reference 1).

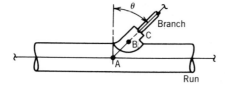

FIGURE 10.6 SIF modeling for 45-degree latrolet.

45° Latrolet, 45° Threaded Latrolet

Member A to B (Fig. 10.6) is modeled using the same cross section as the run pipe but is weightless. Member B to C is modeled using the same internal diameter as the branch pipe but has twice the wall thickness but is weightless. A lumped weight should be added to point B for the latrolet. The weight of any water or insulation will be included on the pipe cross-section card.

End Preparation	Stress Intensification Factor Description	SIF
Butt welded or general	Point A (all connecting members) and Point C (both connecting members)[a] $$SIF = \frac{0.9}{h^{2/3}} \qquad h = \frac{1.97}{r} t$$ t = nominal wall thickness of run pipe r = mean radius of run pipe	Computed
	Point B (both connecting members)	1.0

Sweepolet

In Figure 10.7 member A to B is modeled using the same cross section as the run pipe but is weightless. Member B to C is modeled the same as the branch pipe including weight per foot of pipe.

End Preparation	Stress Intensification Factor Description	SIF
Butt welded or general	Point A (all connecting members)	Computed
	For $r_B/r_R \leq 0.5$ $$SIF_R = 0.8\left(\frac{r_R}{t_R}\right)^{2/3} \times \left(\frac{r_B}{r_R}\right)(F_s)$$ $SIF_R \geq 1.0$	
	For $r_B/r_R > 0.5$ $$SIF_R = 0.4\left(\frac{r_R}{t_R}\right)^{2/3} \times (F_s F_2)$$ $SIF_R \geq 1.5$	
	Point C (both connecting members)[a]	Computed

[a] At the branch pipe member side of point C use the larger of the SIF calculated here and the straight pipe SIF. This applies also to the following point C references.

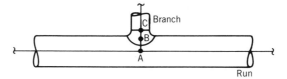

FIGURE 10.7   SIF modeling for sweepolet.

Sockolet, Thredolet, Weldolet

In Figure 10.8 member A to B is modeled using the same cross section as the run pipe but is weightless.

Member B to C (Fig. 10.8) is modeled using the same internal diameter as the branch pipe but has twice the wall thickness but is weightless. A lumped weight should be added to point B for the sockolet. The weight of any water or insulation will be included on the pipe cross-section card.

End Preparation	Stress Intensification Factor Description	SIF
Butt welded or general	Point A (all connecting members)	Computed
	Point C (both connecting members)[a]	
	$SIF = \dfrac{0.9}{h^{2/3}}$ $h = \dfrac{3.3t}{r}$	
	$SIF \geq 1.0$	
	t = nominal wall thickness of the run pipe	
	r = mean radius of the run pipe	
	Point B (both connecting members)	1.0

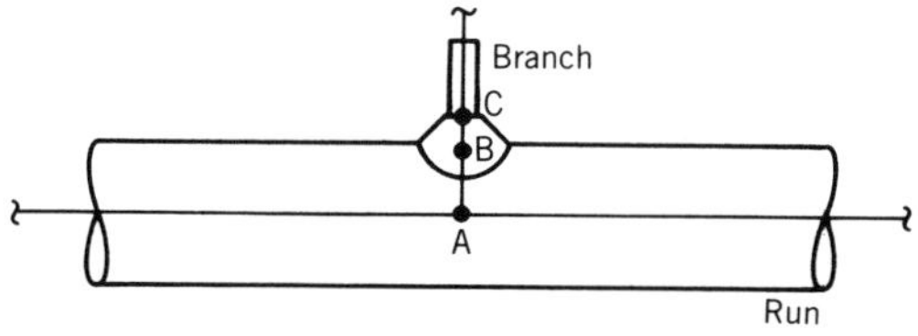

FIGURE 10.8   SIF modeling for sockolet, thredolet, and weldolet.

45° Lateral, 45° Reducing Lateral, 45° Socket Weld Lateral, 45° Threaded Lateral

For butt-welded laterals, members A to B and B to D (Fig. 10.9) are modeled the same as the run pipe including weight per foot of pipe, and member B to C is modeled the same as the branch pipe including weight per foot of pipe.

[a] At the branch pipe member side of point C use the larger of the SIF calculated here and the straight pipe SIF. This applies also to the following point C references.

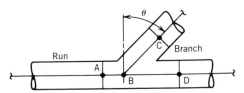

FIGURE 10.9 SIF modeling for 45-degree lateral.

For socket-welded type laterals members A to B, B to C, and B to D are modeled using the same nominal diameter pipe as the run pipe with sch 80 for 3000 lb class fittings and sch 160 used for 6000 lb class fittings but weightless (reference 2). A lumped weight should be added to point B for the lateral. The weight of any water or insulation will be included on the pipe cross-section card.

End Preparation	Stress Intensification Factor Description	SIF
Butt welded or general	Point B	Computed
	$$SIF = \frac{0.9}{h^{2/3}} \qquad h = \frac{t}{r}\,1.97$$	
	$t = $ nominal wall thickness of run pipe	
	$r = $ mean radius of run pipe	
	$\theta = 45°$	
	Points A, C, and D (lateral side)	
	for $t \geq 0.322$	1.0
	$t < 0.322$	1.8
	30° tapered transition	1.9
	$t = $ nominal wall thickness	

Weld Boss, Socket Weld Half Coupling, Threaded Half Coupling, Weld Couplet

Member A to B (Fig. 10.10) is modeled using the same cross section as the run pipe but is weightless. Member B to C is modeled using the same nominal diameter pipe as the branch with sch 80 for 3000 lb class fittings and sch 160 for 6000 lb class fittings but weightless. A lumped weight

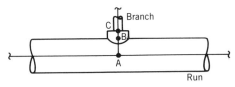

FIGURE 10.10 SIF modeling for weld boss, half coupling, and weld couplet.

should be added to point B for the weld boss. The weight of any water or insulation will be included on the pipe cross-section card.

End Preparation	Stress Intensification Factor Description	SIF
Butt welded or general	Point A (all connecting members) and Point C (both connecting members)[a]	Computed

$$SIF = 1.5\left(\frac{R_m}{T_r}\right)^{2/3}\left(\frac{r'_m}{R_m}\right)^{1/2}\left(\frac{T_b}{T_r}\right)\left(\frac{r'_m}{r_p}\right)$$

$$SIF \geq 1.0 \qquad \frac{R_m}{T_r} \leq 50 \qquad \frac{r'_m}{R_m} \leq 0.5$$

R_m = mean radius of run pipe
r'_m = mean radius of branch pipe
T_r = nominal wall thickness of run pipe
T_b = nominal wall thickness of branch pipe
r_b = outside radius of coupling or boss

| | Point B (both connecting members) | 2.25 |

Tee, Socket Weld Tee, Reducing Tee, Threaded Tee

For butt-welded tees (Figure 10.11), members A to B and B to D are modeled the same as the run pipe including weight per foot of pipe, and member B to C is modeled the same as the branch pipe. For socket-welded type tees, members A to B, B to C, and B to D use the same nominal diameter as the run pipe with sch 80 used for 3000 lb class fittings and sch 160 used for 6000 lb class fittings but weightless (reference 2). A lumped weight should be added to point B for the tee. The weight of any water or insulation will be included on the pipe cross-section card.

End Preparation	Stress Intensification Factor Description	SIF
Butt welded or general	Point B	Computed
	Points A, C, and D (tee side)	
	for $t \geq 0.322$	1.0
	$t < 0.322$	1.8
	30° tapered transition	1.9
	t = nominal wall thickness	
Socket welded	Point B	Computed
	Points A, C, and D (tee side)	1.0

[a] At the branch pipe member side of point C use the larger of the SIF calculated here and the straight pipe SIF. This applies also to the following point C references.

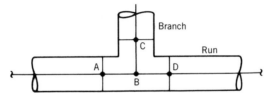

FIGURE 10.11   SIF modeling for tee, socket weld tee, and reducing tee.

Straight Pipe

Member A to B (Fig. 10.12) has the weight of the pipe and any water or insulation on the pipe cross-section card. See other sheets for intensification factors due to branch attachments on pipe.

End Preparation	Stress Instensification Factor Description	SIF
Butt welded	Points A and B (member side)	
or general	for $t \geq 0.322$	1.0
	$t < 0.322$	1.8
	30° tapered transition	1.9
	t = nominal wall thickness of the pipe	
Lap joint flange	Points A and B	1.6
Socket welded	Points A and B	2.1
Threaded	Points A and B	2.3
Slip on flange	Points A and B	2.1

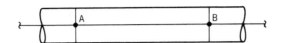

FIGURE 10.12   SIF modeling for straight pipe.

Concentric Reducer

Member A to B (Fig. 10.13) is modeled the same as the largest attached pipe including weight per foot of pipe.

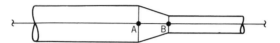

FIGURE 10.13 SIF modeling for concentric reducer.

End Preparation	Stress Intensification Factor Description	SIF
Butt welded or general	Point A (reducer side) Point B (reducer side)	2.0
	$$SIF = 2.0 \frac{Z_1}{Z_2}$$	Computed
	Z_1 = section modulus of the larger pipe Z_2 = section modulus of the smaller pipe	

Eccentric Reducer

Member A to B (Fig. 10.14) is modeled the same as the larger pipe including weight per foot of pipe. The offset between A and B is modeled.

End Preparation	Stress Intensification Factor Description	SIF
Butt welded or general	Point A (reducer side)	Computed
	$$SIF = \frac{0.9}{h^{2/3}} \qquad h = 4.4 \frac{t}{r_1}$$	
	Point B (reducer side)	Computed
	$$SIF = \left(\frac{0.9}{h^{2/3}}\right)\left(\frac{Z_1}{Z_2}\right) \qquad h = 4.4 \frac{t}{r_1}$$	
	$SIF \geq 2.0$	
	t = nominal wall thickness of the larger pipe r_1 = mean radius of the larger pipe Z_1 = section modulus of the larger pipe Z_2 = section modulus of the smaller pipe	

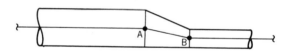

FIGURE 10.14 SIF modeling for ecentric reducer.

Reducing Insert

Member A to B (Fig. 10.15) is modeled using the same nominal diameter pipe as the pipe connected to point B but weightless: sch 80 for 3000 lb class fittings and sch 160 for 6000 lb class fittings (reference 2). A lumped weight should be added to point A for the insert. The weight of any water or insulation will be included on the pipe cross-section card.

SIF = 2.25 (same as socket-welded end).

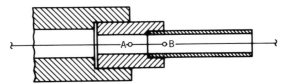

FIGURE 10.15   SIF modeling for reducing insert.

Coupling, Threaded Coupling, Socket Weld Reducer Coupling

Member A to B (Fig. 10.16) is modeled using the same nominal diameter as the coupled pipe (largest nominal diameter if it is a reducing coupling) with sch 80 for 3000 lb class fittings and sch 160 for 6000 lb class fittings but weightless (reference 2). A lumped weight should be added to points A and B for the coupling. The weight of any water or insulation will be included on the pipe cross-section card.

SIF = 2.25 (same as socket-welded end).

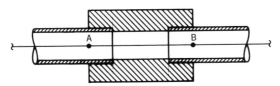

FIGURE 10.16   SIF modeling for coupling.

All Elbows

For butt-welded elbows (Fig. 10.17) member A to C is modeled the same as the attached pipe including weight per foot of pipe. A 180° elbow is modeled as two 90° elbows. If it is a reducing elbow, member A to C is modeled the same as the largest attached pipe.

For socket-welded type elbows, members A to B and B to C are modeled as straight members using the same nominal diameter as the attached pipe with sch 80 for 3000 lb class fittings and sch 160 for 6000 lb class fittings but weightless (reference 2). A lump weight should be added to point B for

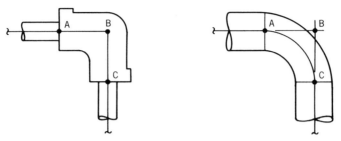

FIGURE 10.17   SIF modeling for elbows.

the elbow. The weight of any water or insulation will be included on the pipe cross-section card.

Elbolet (Socket Weld, Butt Weld, Threaded)

Member A to C (Fig. 10.18) is modeled using the same cross section as the run pipe but is weightless. Member C to D is modeled using the same internal diameter as the branch pipe but has twice the wall thickness but weightless. A lumped weight should be added to point C for the elbolet. The weight of any water or insulation will be included on the pipe cross-section card.

End Preparation	Stress Intensification Factor Description	SIF
Butt welded or general	Point A (member A to B) Same as the elbow	
	Point A (member A to C) and point C both connecting members)	1.0
	Point D (both connecting members)[a]	Computed

$$S1 = \frac{0.9}{h_1^{2/3}} \qquad \begin{matrix} S1 \geq 1.0 \\ S2 \geq 1.0 \end{matrix}$$

$$S2 = \frac{0.9}{h_2^{2/3}} \qquad SIF = S1 \times S2$$

$$h_1 = \frac{t}{2r} \qquad h_2 = \frac{tR}{r^2}$$

t = nominal wall thickness of the run
        pipe

r = mean radius of the run pipe

R = bend radius of elbow

---

[a] At the branch pipe member side of point C use the larger of the SIF calculated here and the straight pipe SIF. This applies also to the following point C references.

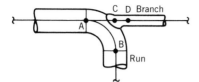

FIGURE 10.18 SIF modeling for elbolet.

Valves, Valve with No Operator

In Figure 10.19, members A to B, B to C, and B to D (if operator exists) are modeled using the same internal diameter as the attached pipe but with twice the wall thickness but weightless. Members A to B and B to C have the weight of any water or insulation on the pipe cross-section card. Member B to D is weightless. Lumped weights of the valve and operator (if operator exists) should be added to the points where needed. Two mass points, one for valve and one for operator C.G, are required.

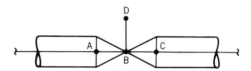

FIGURE 10.19 SIF modeling for valves with no operator.

All Flanges

Members A to B and B to C (Fig. 10.20) are modeled using the same internal diameter as the attached pipe but with twice and wall thickness but weightless. A lumped weight should be added to B for the flange(s). The weight of any water or insulation will be on the pipe cross-section card.

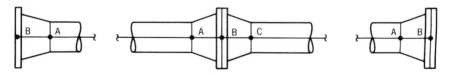

FIGURE 10.20 SIF modeling for flanges.

Cap

Member A to B (Fig. 10.21) is modeled using the same cross sections as the run pipe but is weightless. A lumped weight should be added to point B

FIGURE 10.21   SIF modeling for caps.

for the cap. The weight of any water on insulation will be included on the pipe cross-section card.

REFERENCES

1. Tennessee Valley Authority, *Piping Analysis Procedure*.
2. *Oakridge National Laboratory Report ORNL-TM-4929*.

APPENDIXES

TABLE A1 Total Thermal Expansion, U.S. Units, for Metals
Total Linear Thermal Expansion Between 70°F and Indicated Temperature (inches/100 ft)[a]

Temp. °F	Carbon Steel Carbon-Moly Low-Chrome (thru 3 Cr Mo)	5 Cr Mo thru 9 Cr Mo	MATERIAL Austenitic Stainless Steels 18 Cr–8 Ni	12 Cr 17 Cr 27 Cr	25 Cr–20 Ni	Monel 67 Ni–30 Cu	3½ Nickel
−325	−2.37	−2.22	−3.85	−2.04	—	−2.62	−2.25
−300	−2.24	−2.10	−3.63	−1.92	—	−2.50	−2.17
−275	−2.11	−1.98	−3.41	−1.80	—	−2.38	−2.07
−250	−1.98	−1.86	−3.19	−1.68	—	−2.26	−1.96
−225	−1.85	−1.74	−2.96	−1.57	—	−2.14	−1.86
−200	−1.71	−1.62	−2.73	−1.46	—	−2.02	−1.76
−175	−1.58	−1.50	−2.50	−1.35	—	−1.90	−1.62
−150	−1.45	−1.37	−2.27	−1.24	—	−1.79	−1.48
−125	−1.30	−1.23	−2.01	−1.11	—	−1.59	−1.33
−100	−1.15	−1.08	−1.75	−0.98	—	−1.38	−1.17
−75	−1.00	−0.94	−1.50	−0.85	—	−1.18	−1.01
−50	−0.84	−0.79	−1.24	−0.72	—	−0.98	−0.84
−25	−0.68	−0.63	−0.98	−0.57	—	−0.77	−0.67
0	−0.49	−0.46	−0.72	−0.42	—	−0.57	−0.50
25	−0.32	−0.30	−0.46	−0.27	—	−0.37	−0.32
50	−0.14	−0.13	−0.21	−0.12	—	−0.20	−0.15
70	0	0	0	0	0	0	0

Temperature							
100	0.23	0.22	0.34	0.20	0.32	0.28	0.23
125	0.42	0.40	0.62	0.36	0.58	0.52	0.42
150	0.61	0.58	0.90	0.53	0.84	0.75	0.61
175	0.80	0.76	1.18	0.69	1.10	0.99	0.81
200	0.99	0.94	1.46	0.86	1.37	1.22	1.01
225	1.21	1.13	1.75	1.03	1.64	1.46	1.21
250	1.40	1.33	2.03	1.21	1.91	1.71	1.42
275	1.61	1.52	2.32	1.38	2.18	1.96	1.63
300	1.82	1.71	2.61	1.56	2.45	2.21	1.84
325	2.04	1.90	2.90	1.74	2.72	2.44	2.05
350	2.26	2.10	3.20	1.93	2.99	2.68	2.26
375	2.48	2.30	3.50	2.11	3.26	2.91	2.47
400	2.70	2.50	3.80	2.30	3.53	3.25	2.69
425	2.93	2.72	4.10	2.50	3.80	3.52	2.91
450	3.16	2.93	4.41	2.69	4.07	3.79	3.13
475	3.39	3.14	4.71	2.89	4.34	4.06	3.35
500	3.62	3.35	5.01	3.08	4.61	4.33	3.58
525	3.86	3.58	5.31	3.28	4.88	4.61	3.81
550	4.11	3.80	5.62	3.49	5.15	4.90	4.04
575	4.35	4.02	5.93	3.69	5.42	5.18	4.27
600	4.60	4.24	6.24	3.90	5.69	5.46	4.50
625	4.86	4.47	6.55	4.10	5.96	5.75	4.74
650	5.11	4.69	6.87	4.31	6.23	6.05	4.98

[a]These data are for information and it is not implied that materials are suitable for all the temperatures shown.

TABLE A1 (Continued)
Total Linear Thermal Expansion Between 70°F and Indicated Temperature (inches/100 ft)[a]

Temp. °F	Carbon Steel Carbon-Moly Low-Chrome (thru 3 Cr Mo)	5 Cr Mo thru 9 Cr Mo	MATERIAL Austenitic Stainless Steels 18 Cr-8 Ni	12 Cr 17 Cr 27 Cr	25 Cr-20 Ni	Monel 67 Ni-30 Cu	3½ Nickel
675	5.37	4.92	7.18	4.52	6.50	6.34	5.22
700	5.63	5.14	7.50	4.73	6.77	6.64	5.46
725	5.90	5.38	7.82	4.94	7.04	6.94	5.70
750	6.16	5.62	8.15	5.16	7.31	7.25	5.94
775	6.43	5.86	8.47	5.38	7.58	7.55	6.18
800	6.70	6.10	8.80	5.60	7.85	7.85	6.43
825	6.97	6.34	9.13	5.82	8.15	8.16	6.68
850	7.25	6.59	9.46	6.05	8.45	8.48	6.93
875	7.53	6.83	9.79	6.27	8.75	8.80	7.18
900	7.81	7.07	10.12	6.49	9.05	9.12	7.43
925	8.08	7.31	10.46	6.71	9.35	9.44	7.68
950	8.35	7.56	10.80	6.94	9.65	9.77	7.93
975	8.62	7.81	11.14	7.17	9.95	10.09	8.17
1000	8.89	8.06	11.48	7.40	10.25	10.42	8.41

1025	9.17	8.30	11.82	7.62	10.55	10.75	–
1050	9.46	8.55	12.16	7.95	10.85	11.09	–
1075	9.75	8.80	12.50	8.18	11.15	11.43	–
1100	10.04	9.05	12.84	8.31	11.45	11.77	–
1125	10.31	9.28	13.18	8.53	11.78	12.11	–
1150	10.57	9.52	13.52	8.76	12.11	12.47	–
1175	10.83	9.76	13.86	8.98	12.44	12.81	–
1200	11.10	10.00	14.20	9.20	12.77	13.15	–
1225	11.38	10.26	14.54	9.42	13.10	13.50	–
1250	11.66	10.53	14.88	9.65	13.43	13.86	–
1275	11.94	10.79	15.22	9.88	13.76	14.22	–
1300	12.22	11.06	15.56	10.11	14.09	14.58	–
1325	12.50	11.30	15.90	10.33	14.39	14.49	–
1350	12.78	11.55	16.24	10.56	14.69	15.30	–
1375	13.06	11.80	16.58	10.78	14.99	15.66	–
1400	13.34	12.05	16.92	11.01	15.29	16.02	–
1425	–	–	17.30	–	–	–	–
1450	–	–	17.69	–	–	–	–
1475	–	–	18.08	–	–	–	–
1500	–	–	18.47	–	–	–	–

[a]These data are for information and it is not implied that materials are suitable for all the temperatures shown.

TABLE A1 (Continued)
Total Linear Thermal Expansion Between 70°F and Indicated Temperature (inches/100 ft)[a]

MATERIAL

Aluminum	Gray Cast Iron	Bronze	Brass	70 Cu-30 Ni	Ni-Fe-Cr	Ni-Cr-Fe	Ductile Iron	Temp. °F
-4.68	—	-3.98	-3.88	-3.15	—	—	—	-325
-4.46	—	-3.74	-3.64	-2.87	—	—	—	-300
-4.21	—	-3.50	-3.40	-2.70	—	—	—	-275
-3.97	—	-3.26	-3.16	-2.53	—	—	—	-250
-3.71	—	-3.02	-2.93	-2.36	—	—	—	-225
-3.44	—	-2.78	-2.70	-2.19	—	—	-1.51	-200
-3.16	—	-2.54	-2.47	-2.12	—	—	-1.41	-175
-2.88	—	-2.31	-2.24	-1.95	—	—	-1.29	-150
-2.57	—	-2.06	-2.00	-1.74	—	—	-1.16	-125
-2.27	—	-1.81	-1.76	-1.53	—	—	-1.04	-100
-1.97	—	-1.56	-1.52	-1.33	—	—	-0.91	-75
-1.67	—	-1.32	-1.29	-1.13	—	—	-0.77	-50
-1.32	—	-1.25	-1.02	-0.89	—	—	-0.62	-25
-0.97	—	-0.77	-0.75	-0.66	—	—	-0.46	0
-0.63	—	-0.49	-0.48	-0.42	—	—	-0.23	25
-0.28	—	-0.22	-0.21	-0.19	—	—	-0.14	50
0	0	0	0	0	0	0	0.0	70
0.46	0.21	0.36	0.35	0.31	0.28	0.26	0.21	100

206

Temp								
125	0.39	0.48	0.52	0.56	0.64	0.66	0.85	0.38
150	0.57	0.70	0.76	0.82	0.94	0.96	1.23	0.55
175	0.76	0.92	0.99	1.07	1.23	1.26	1.62	0.73
200	0.94	1.15	1.23	1.33	1.52	1.56	2.00	0.90
225	1.13	1.38	1.49	1.59	1.83	1.86	2.41	1.08
250	1.33	1.61	1.76	1.86	2.14	2.17	2.83	1.27
275	1.53	1.85	2.03	2.13	2.45	2.48	3.24	1.45
300	1.72	2.09	2.30	2.40	2.76	2.79	3.67	1.64
325	1.93	2.32	2.59	2.68	3.08	3.11	4.09	1.83
350	2.13	2.56	2.88	2.96	3.41	3.42	4.52	2.03
375	2.36	2.80	3.18	3.24	3.73	3.74	4.95	2.22
400	2.56	3.05	3.48	3.52	4.05	4.05	5.39	2.42
425	2.79	3.29	3.76	–	4.38	4.37	5.83	2.62
450	3.04	3.53	4.04	–	4.72	4.69	6.28	2.83
475	3.28	3.78	4.31	–	5.06	5.01	6.72	3.03
500	3.54	4.02	4.59	–	5.40	5.33	7.17	3.24
525	3.76	4.27	4.87	–	5.75	5.65	7.63	3.46
550	3.99	4.52	5.16	–	6.10	5.98	8.10	3.67
575	4.22	4.77	5.44	–	6.45	6.31	8.56	3.89
600	4.44	5.02	5.72	–	6.80	6.64	9.03	4.11
625	4.66	5.27	6.01	–	7.16	6.96	–	4.34
650	4.90	5.53	6.30	–	7.53	7.29	–	4.57

[a]These data are for information and it is not implied that materials are suitable for all the temperatures shown.

TABLE A1 (Continued)
Total Linear Thermal Expansion Between 70°F and Indicated Temperature (inches/100 ft)[a]

				Material				
Aluminum	Gray Cast Iron	Bronze	Brass	70 Cu–30 Ni	Ni–Fe–Cr	Ni–Cr–Fe	Ductile Iron	Temp. °F
–	4.80	7.62	7.89	–	6.58	5.79	5.14	675
–	5.03	7.95	8.26	–	6.88	6.05	5.39	700
–	5.26	8.28	8.64	–	7.17	6.31	5.60	725
–	5.50	8.62	9.02	–	7.47	6.57	5.85	750
–	5.74	8.96	9.40	–	7.76	6.84	6.10	775
–	5.98	9.30	9.78	–	8.06	7.10	6.35	800
–	6.22	9.64	10.17	–	8.35	–	6.59	825
–	6.47	9.99	10.57	–	8.66	–	6.85	850
–	6.72	10.33	10.96	–	8.95	–	7.09	875
–	6.97	10.68	11.35	–	9.26	–	7.35	900
–	7.23	11.02	11.75	–	9.56	–	7.64	925
–	7.50	11.37	12.16	–	9.87	–	7.86	950
–	7.76	11.71	12.57	–	10.18	–	8.11	975
–	8.02	12.05	12.98	–	10.49	–	8.35	1000
–	–	12.40	13.39	–	10.80	–	–	1025

Temperature							
1050	—	11.11	—	13.81	12.76	—	—	—
1075	—	11.42	—	14.23	13.11	—	—	—
1100	—	11.74	—	14.65	13.47	—	—	—
1125	—	12.05	—	—	—	—	—	—
1150	—	12.38	—	—	—	—	—	—
1175	—	12.69	—	—	—	—	—	—
1200	—	13.02	—	—	—	—	—	—
1225	—	13.36	—	—	—	—	—	—
1250	—	13.71	—	—	—	—	—	—
1275	—	14.04	—	—	—	—	—	—
1300	—	14.39	—	—	—	—	—	—
1325	—	14.74	—	—	—	—	—	—
1350	—	15.10	—	—	—	—	—	—
1375	—	15.44	—	—	—	—	—	—
1400	—	15.80	—	—	—	—	—	—
1425	—	16.16	—	—	—	—	—	—
1450	—	16.53	—	—	—	—	—	—
1475	—	16.88	—	—	—	—	—	—
1500	—	17.25	—	—	—	—	—	—

[a]These data are for information and it is not implied that materials are suitable for all the temperatures shown.

TABLE A2 Modulus of Elasticity, U.S. Units, for Metals
Modulus of Elasticity—Ferrous Material[a]

E = Modulus of Elasticity, ksi (multiply tabulated values by 10^3)

Material	\-325	\-200	\-100	70	200	300	400	500	600	700	800	900	1000	1100	1200	1300	1400	1500
										Temperature, °F								
Carbon steels with carbon content 0.30% or less, 3½ Ni	30.0	29.5	29.0	27.9	27.7	27.4	27.0	26.4	25.7	24.8	23.4	18.5	15.4	13.0	–	–	–	–
Carbon steels with carbon content above 0.30%	31.0	30.6	30.4	29.9	29.5	29.0	28.3	27.4	26.7	25.4	23.8	21.5	18.8	15.0	11.2	–	–	–
Carbon-Moly steels, low chrome steels through 3 Cr Mo	31.0	30.6	30.4	29.9	29.5	29.0	28.6	28.0	27.4	26.6	25.7	24.5	23.0	20.4	15.6	–	–	–
Intermediate chrome steels (5 Cr Mo through 9 Cr Mo)	29.4	28.5	28.1	27.4	27.1	26.8	26.4	26.0	25.4	24.9	24.2	23.5	22.8	21.9	20.8	19.5	18.1	–
Austenitic steels (TP304, 310, 316, 321, 347)	30.4	29.9	29.4	28.3	27.7	27.1	26.6	26.1	25.4	24.8	24.1	23.4	22.7	22.0	21.3	20.7	19.3	17.9

(table continued from previous page)

Material	–325	–200	–100	70	100	200	300	400	500	600	700	800	900	1000	1100	1200
Straight chromium steels (12 Cr, 17 Cr, 27 Cr)	30.8	30.3	29.8	29.2	28.7	28.3	27.7	27.0	26.0	24.8	23.1	21.1	18.6	15.6	12.2	–
Gray cast iron	–	–	–	13.4	13.2	12.9	12.6	12.2	11.7	11.0	10.2	–	–	–	–	–

Modulus of Elasticity—Nonferrous Materials

E = Modulus of Elasticity, ksi (multiply tabulated values by 10^3)

Material	Temperature, °F															
	–325	–200	–100	70	100	200	300	400	500	600	700	800	900	1000	1100	1200
Monel (67 Ni–30 Cu) and 66 Ni–29 Cu–Al	26.8	26.6	26.4	26.0	26.0	26.0	25.8	25.6	25.4	24.7	23.1	21.0	18.6	16.0	14.3	13.0
Copper–Nickel (70 Cu–30 Ni)	–	–	–	21.6	21.5	21.2	20.9	20.6	20.3	20.0	19.7	19.4	–	–	–	–
Aluminum alloys	11.3	10.9	10.6	10.1	10.0	9.8	9.5	8.7	7.7	–	–	–	–	–	–	–
Copper (99.98% Cu)	17.0	16.7	16.5	16.0	15.8	15.6	15.4	15.1	14.7	14.2	13.7	–	–	–	–	–
Commercial brass (66 Cu –34 Zn)	15.0	14.7	14.5	14.0	13.9	13.7	13.5	13.0	12.7	12.2	11.8	–	–	–	–	–
Leaded tin bronze (88 Cu–6 Sn–1.5 Pb–4.5 Zn)	14.2	13.8	13.5	13.0	12.9	12.7	12.4	12.0	11.7	11.3	10.9	–	–	–	–	–

[a]These data are for information and it is not implied that materials are suitable for all the temperatures shown.

TABLE A3 Allowable Stresses in Tension for Metals, SE, KSI

Material	Speci-fication	P-No. (37)	Grade	Class	Factor (E)	Min. Tensile Strength (ksi)	Min. Yield Strength (ksi)	Notes	Min. Temp. (26)	Min. Temp. to 100	200	300
Carbon Steel												
Seamless Pipe and Tubes												
–	A53	1	A	Type S	–	48.0	30.0	1, 2	–20	16.0	16.0	16.0
–	A53	1	B	Type S	–	60.0	35.0	1, 2	–20	20.0	20.0	20.0
–	A106	1	A	–	–	48.0	30.0	2	–20	16.0	16.0	16.0
–	A106	1	B	–	–	60.0	30.0	2	–20	20.0	20.0	20.0
–	A106	1	C	–	–	70.0	40.0	2	–20	23.3	23.3	23.3
–	A120	1	–	–	–	–	–	34	–20	12.0	11.4	–
–	A333	1	1	–	–	55.0	30.0	1, 2	–50	18.3	18.3	17.7
–	A334	1	1	–	–	55.0	30.0	1, 2	–50	18.3	18.3	17.7
–	A333	1	6	–	–	60.0	35.0	2	–50	20.0	20.0	20.0
–	A334	1	6	–	–	60.0	35.0	2	–50	20.0	20.0	20.0
–	A179	1	–	–	–	47.0	26.0	1, 2	–20	15.7	15.0	14.2
–	A369	1	FPA	–	–	48.0	30.0	2	–20	16.0	16.0	16.0
–	A369	1	FPB	–	–	60.0	35.0	2	–20	20.0	20.0	20.0
–	A524	1	I	–	–	60.0	35.0	2	–20	20.0	20.0	20.0
–	A524	1	II	–	–	55.0	30.0	2	–20	18.3	18.3	17.7
–	API 5L	1	A	–	–	48.0	30.0	1, 2	–20	16.0	16.0	16.0
–	API 5L	1	B	–	–	60.0	35.0	1, 2	–20	20.0	20.0	20.0
–	API 5LX	SP2	X42	–	–	60.0	42.0	58, 60	–20	20.0	20.0	20.0
–	API 5LX	SP3	X46	–	–	63.0	46.0	58, 60	–20	21.0	21.0	21.0
–	API 5LX	SP3	X52	–	–	66.0	52.0	58, 60	–20	22.0	22.0	22.0

–	API 5LX	SP3	X52	–	–	72.0	52.0	58, 60	–20	24.0	24.0	24.0
Furnace Butt Welded Pipe												
–	A53	1	–	Type F	0.60	45.0	25.0	34	–20	‖9.0	9.0	8.7
–	A120	1	–	–	0.60	–	–	34	–20	‖7.2	6.9	–
–	API 5L	1	A25	I & II	0.60	45.0	25.0	34	–20	‖9.0	9.0	8.7
Electric Resistance Welded Pipe												
–	A53	1	A	Type E	0.85	48.0	30.0	1, 2	–20	13.6	13.6	13.6
–	A53	1	B	Type E	0.85	60.0	35.0	1, 2	–20	17.0	17.0	17.0
–	A120	1	–	–	0.85	–	–	34	–20	10.2	9.7	–
–	A135	1	A	–	0.85	48.0	30.0	1, 2	–20	13.6	13.6	13.6
–	A135	1	B	–	0.85	60.0	35.0	1, 2	–20	17.0	17.0	17.0
–	A333	1	1	–	0.85	55.0	30.0	1, 2	–50	15.6	15.6	15.0
–	A333	1	6	–	0.85	60.0	35.0	2	–50	17.0	17.0	17.0
–	A587	1	–	–	0.85	48.0	30.0	1, 2	–20	13.6	13.6	13.6
–	API 5L	1	A25	I & II	0.85	45.0	25.0	1, 2	–20	12.8	12.8	12.3
–	API 5L	1	A	–	0.85	48.0	30.0	1, 2	–20	13.6	13.6	13.6
–	API 5L	1	B	–	0.85	60.0	35.0	1, 2	–20	17.0	17.0	17.0
–	API 5L	SP2	X42	–	0.85	60.0	42.0	58, 60	–20	17.0	17.0	17.0
–	API 5LX	SP3	X46	–	0.85	63.0	46.0	58, 60	–20	17.9	17.9	17.9
–	API 5LX	SP3	X52	–	0.85	66.0	52.0	58, 60	–20	18.7	18.7	18.7
–	API 5LX	SP3	X52	–	0.85	72.0	52.0	58, 60	–20	20.4	20.4	20.4
Electric Fusion Welded Pipe (Straight Seam)												
A570 GR A	A134	1	–	–	0.74	45.0	25.0	5, 34	–20	11.1	10.5	10.0
A570 GR B	A134	1	–	–	0.74	49.0	30.0	5, 34	–20	12.1	11.4	10.9
A570 GR C	A134	1	–	–	0.74	52.0	33.0	5, 34	–20	12.8	12.1	11.6
A570 GR D	A134	1	–	–	0.74	55.0	40.0	5, 34	–20	13.6	12.8	12.2

[a]See B31.3 code for notes.

TABLE A3 (Continued)
Allowable Stresses in Tension for Metals, SE, KSI

400	500	600	650	700	750	800	850	900	950	1000	1050	1100	1150	Specification	Material
															Carbon Steel Seamless Pipe and Tubes
16.0	16.0	14.8	14.5	14.4	10.7	9.3	7.9	6.5	4.5	2.5	1.6	1.0	–	A53	–
20.0	18.9	17.3	17.0	16.8	13.0	10.8	8.7	6.5	4.5	2.5	1.6	1.0	–	A53	–
16.0	16.0	14.8	14.5	14.4	10.7	9.3	7.9	6.5	4.5	2.5	1.6	1.0	–	A106	–
20.0	18.9	17.3	17.0	16.8	13.0	10.8	8.7	6.5	4.5	2.5	1.6	1.0	–	A106	–
22.9	21.6	19.7	19.4	19.2	14.8	12.0|	–	–	–	–	–	–	–	A106	–
–	–	–	–	–	–	–	–	–	–	–	–	–	–	A120	–
17.2	16.2	14.8	14.5	14.4	12.0	10.2	8.3	6.5	4.5	2.5	1.6	1.0	–	A333	–
17.2	16.2	14.8	14.5	14.4	12.0	10.2	8.3	6.5	4.5	2.5	1.6	1.0	–	A334	–
20.0	18.9	17.3	17.0	16.8	13.0	10.8	8.7	6.5	4.5	2.5	1.6	1.0	–	A333	–
20.0	18.9	17.3	17.0	16.8	13.0	10.8	8.7	6.5	4.5	2.5	1.6	1.0	–	A334	–
13.5	12.8	12.1	11.8	11.5	10.6	9.2	7.9	6.5	4.5	2.5	1.6	1.0	–	A179	–
16.0	16.0	14.8	14.5	14.4	10.7	9.3|	7.9	6.5	4.5	2.5	1.6	1.0	–	A369	–
20.0	18.9	17.3	17.0	16.8	13.0	10.8|	8.7	6.5	4.5	2.5	1.6	1.0	–	A369	–
20.0	18.9	17.3	17.0	16.8	12.9	10.8	8.6	6.5	4.5	2.5	–	–	–	A524	–
17.2	16.2	14.8	14.6	14.4	12.0	10.2	8.3	6.5	4.5	2.5	–	–	–	A524	–
16.0	16.0	14.8	14.5	14.4	10.7	9.3|	7.9	6.5	4.5	2.5	1.6	1.0	–	API 5L	–
20.0	18.9	17.3	17.0	16.8	13.0	10.8|	8.7	6.5	4.5	2.5	1.6	1.0	–	API 5L	–
20.0	–	–	–	–	–	–	–	–	–	–	–	–	–	API 5LX	–
21.0	–	–	–	–	–	–	–	–	–	–	–	–	–	API 5LX	–
22.0	–	–	–	–	–	–	–	–	–	–	–	–	–	API 5LX	–
24.0	–	–	–	–	–	–	–	–	–	–	–	–	–	API 5LX	–

Type / Spec														
Furnace Butt-Welded Pipe														
A53	—	—	—	—	—	—	—	—	—	—	—	—	—	8.3
A120	—	—	—	—	—	—	—	—	—	—	—	—	—	—
API 5L	—	—	—	—	—	—	—	—	—	—	—	—	—	8.3
Electric Resistance-Welded Pipe														
A53	—	0.9	1.4	2.1	3.8	5.5	6.7	7.9	9.1	12.2	12.3	12.6	13.6	13.6
A53	—	0.9	1.4	2.1	3.8	5.5	7.4	9.2	11.0	14.0	14.5	14.7	16.1	17.0
A120	—	—	—	—	—	—	—	—	—	—	—	—	—	—
A135	—	0.9	1.4	2.1	3.8	5.5	6.5	7.9	9.1	12.2	12.3	12.6	13.6	13.6
A135	—	—	—	2.1	3.8	5.5	7.4	9.2	11.0	14.0	14.5	14.7	16.1	17.0
A333	—	0.9	1.4	2.1	3.8	5.5	7.1	8.7	10.2	12.2	12.3	12.6	13.8	14.6
A333	—	0.9	1.4	2.1	3.8	5.5	7.4	9.2	11.0	14.0	14.5	14.7	16.1	17.0
A587	—	—	—	—	—	—	—	—	—	—	12.3	12.6	13.6	13.6
API 5L	—	—	—	—	—	—	—	—	—	—	—	—	—	11.8
API 5L	—	0.9	1.4	2.1	3.8	5.5	6.7	7.9	9.1	12.2	12.3	12.6	13.6	13.6
API 5L	—	0.9	1.4	2.1	3.8	5.5	7.4	9.2	11.0	14.0	14.5	14.7	16.1	17.0
API 5LX	—	—	—	—	—	—	—	—	—	—	—	—	—	17.0
API 5LX	—	—	—	—	—	—	—	—	—	—	—	—	—	17.9
API 5LX	—	—	—	—	—	—	—	—	—	—	—	—	—	18.7
API 5LX	—	—	—	—	—	—	—	—	—	—	—	—	—	20.4
Electric Fusion-Welded Pipe (Straight Seam)														
A134 A570 GR A	—	—	—	—	—	—	—	—	—	—	—	—	—	—
A134 A570 GR B	—	—	—	—	—	—	—	—	—	—	—	—	—	—
A134 A570 GR C	—	—	—	—	—	—	—	—	—	—	—	—	—	—
A134 A570 GR D	—	—	—	—	—	—	—	—	—	—	—	—	—	—

215

TABLE A3 (Continued)
Allowable Stresses in Tension for Metals, SE, KSI

Material	Speci-fication	P-No. (37)	Grade	Class	Factor (E)	Min. Tensile Str. (ksi)	Min. Yield Str. (ksi)	Notes	Min. Temp. (26)	Min. Temp. to 100	200
Low and Intermediate Alloy Steel											
Seamless Pipe											
$3\frac{1}{2}$ Ni	A333	9B	3	–	–	65.0	35.0	–	−150	21.7	19.6
$3\frac{1}{2}$ Ni	A334	9B	3	–	–	65.0	35.0	–	−150	21.7	19.6
$\frac{3}{4}$ Cr-$\frac{3}{4}$ Ni-Cu-Al	A333	4	4	–	–	60.0	35.0	–	−150	20.0	19.1
$2\frac{1}{4}$ Ni	A333	9A	7	–	–	65.0	35.0	–	−100	21.7	19.6
$2\frac{1}{4}$ Ni	A334	9A	7	–	–	65.0	35.0	–	−100	21.7	19.6
9 Ni	A333	11A-SG1	8	–	–	100.0	75.0	69	−320	31.7	31.7
9 Ni	A334	11A-SG1	8	–	–	100.0	75.0	69	−320	31.7	31.7
2 Ni-1 Cu	A333	9A	9	–	–	63.0	46.0	–	−100	21.0	–
2 Ni-1 Cu	A334	9A	9	–	–	63.0	46.0	–	−100	21.0	–
Cr-$\frac{1}{2}$ Mo	A335	3	P1	–	–	55.0	30.0	3	−20	18.3	18.3
$\frac{3}{4}$ Cr-$\frac{1}{2}$ Mo	A335	3	P2	–	–	55.0	30.0	–	−20	18.3	18.3
5 Cr-$\frac{1}{2}$ Mo	A335	5	P5	–	–	60.0	30.0	–	−20	20.0	18.1
5 Cr-$\frac{1}{2}$ Mo-Si	A335	5	P5b	–	–	60.0	30.0	–	−20	20.0	18.1
5 Cr-$\frac{1}{2}$ Mo-Ti	A335	5	P5c	–	–	60.0	30.0	–	−20	20.0	18.1
7 Cr-$\frac{1}{2}$ Mo	A335	5	P7	–	–	60.0	30.0	–	−20	20.0	18.1
9 Cr-1 Mo	A335	5	P9	–	–	60.0	30.0	–	−20	20.0	18.1
$1\frac{1}{4}$ Cr-$\frac{1}{2}$ Mo	A335	4	P11	–	–	60.0	30.0	–	−20	20.0	18.7
1 Cr-$\frac{1}{2}$ Mo	A335	4	P12	–	–	60.0	30.0	–	−20	20.0	18.7

Material	Spec	No.	Grade								
1½Si-½Mo	A335	3	P15	–	–	60.0	30.0	–	-20	18.8	18.2
3Cr-1Mo	A335	5	P21	–	–	60.0	30.0	–	-20	20.0	18.7
2¼Cr-1Mo	A335	5	P22	–	–	60.0	30.0	–	-20	20.0	18.5
Cr-½Mo	A369	3	FP1	–	–	55.0	30.0	3	-20	18.3	18.3
¾Cr-½Mo	A369	3	FP2	–	–	55.0	30.0	–	-20	18.3	18.3
2Cr-½Mo	A369	4	FP3b	–	–	60.0	20.0	–	-20	20.0	18.5
5Cr-½Mo	A369	5	FP5	–	–	60.0	30.0	–	-20	20.0	18.1
7Cr-½Mo	A369	5	FP7	–	–	60.0	30.0	–	-20	20.0	18.1
9Cr-1Mo	A369	5	FP9	–	–	60.0	30.0	–	-20	20.0	18.1
1¼Cr-½Mo	A369	4	FP11	–	–	60.0	30.0	–	-20	20.0	18.7
1Cr-½Mo	A369	4	FP12	–	–	60.0	30.0	–	-20	20.0	18.7
3Cr-1Mo	A369	5	FP21	–	–	60.0	30.0	–	-20	20.0	18.7
2¼Cr-1Mo	A369	5	FP22	–	–	60.0	30.0	–	-20	20.0	18.5
Centrifugally Cast Pipe											
C-½Mo	A426	3	CP1	–	1.00	65.0	35.0	3, 21	-20	21.7	21.7
¾Cr-½Mo	A426	3	CP2	–	1.00	60.0	30.0	21	-20	18.4	17.7
5Cr-½Mo	A426	5	CP5	–	1.00	90.0	60.0	21	-20	30.0	28.0
5Cr-½Mo-1½Si	A426	5	CP5b	–	1.00	60.0	30.0	21	-20	18.8	17.9
7Cr-½Mo	A426	5	CP7	–	1.00	60.0	30.0	21	-20	18.8	17.9
9Cr-1Mo	A426	5	CP9	–	1.00	90.0	60.0	21	-20	30.0	22.5
1¼Cr-½Mo	A426	4	CP11	–	1.00	70.0	40.0	21	-21	23.3	23.3
1Cr-½Mo	A426	4	CP12	–	1.00	60.0	30.0	21	-20	18.8	18.3
1½Si-½Mo	A426	3	CP15	–	1.00	60.0	30.0	21	-21	18.8	18.2
12¾Cr	A426	7	CPCA15	–	1.00	90.0	65.0	21	-20	30.0	–
3Cr-1Mo	A426	5	CP21	–	1.00	60.0	30.0	21	-20	18.8	18.1
2¼Cr-1Mo	A426	5	CP22	–	1.00	70.0	40.0	21	-20	23.3	23.3

TABLE A3 (Continued)
Allowable Stresses in Tension for Metals, SE, KSI

300	400	500	600	650	700	750	800	850	900	950	1000	1050	1100	1150	1200	Specification	Material
																	Low and Intermediate Alloy Steel Seamless Pipe
19.6	18.7	17.8	16.8	16.3	15.5	13.9	11.4	9.0	6.5	4.5	2.5	1.6	1.0	–	–	A333	$3\frac{1}{2}$ Ni
19.6	18.7	17.8	16.8	16.3	15.5	13.9	11.4	9.0	6.5	4.5	2.5	1.6	1.0	–	–	A334	$3\frac{1}{2}$ Ni
18.2	17.3	16.4	15.5	15.0	–	–	–	–	–	–	–	–	–	–	–	A333	$\frac{3}{4}$ Cr–$\frac{3}{4}$ Ni–Cu–Al
19.6	18.7	17.6	16.8	16.3	15.5	13.9	11.4	9.0	6.5	4.5	2.5	1.6	1.0	–	–	A333	$2\frac{1}{4}$ Ni
19.6	18.7	17.6	16.8	16.3	15.5	13.9	11.4	9.0	6.5	4.5	2.5	1.6	1.0	–	–	A334	$2\frac{1}{4}$ Ni
–	–	–	–	–	–	–	–	–	–	–	–	–	–	–	–	A333	9 Ni
–	–	–	–	–	–	–	–	–	–	–	–	–	–	–	–	A334	9 Ni
–	–	–	–	–	–	–	–	–	–	–	–	–	–	–	–	A333	2 Ni – 1 Cu
–	–	–	–	–	–	–	–	–	–	–	–	–	–	–	–	A334	2 Ni–1 Cu
17.5	16.9	16.3	15.7	15.4	15.1	13.8	13.5	13.1	12.7	8.2	4.8	–	–	–	–	A335	C–$\frac{1}{2}$ Mo
17.5	16.9	16.3	15.7	15.4	15.1	13.8	13.5	13.1	12.8	9.2	5.9	–	–	–	–	A335	$\frac{3}{4}$ Cr–$\frac{1}{2}$ Mo
17.4	17.2	17.1	16.8	16.6	16.3	13.2	12.8	12.1	10.9	8.0	5.8	4.2	2.9	2.0	1.3	A335	5 Cr–$\frac{1}{2}$ Mo
17.4	17.2	17.1	16.8	16.6	16.3	13.2	12.8	12.1	10.9	8.0	5.8	4.2	2.9	2.0	1.3	A335	5 Cr–$\frac{1}{2}$ Mo-Si
17.4	17.2	17.1	16.8	16.6	16.3	13.2	12.8	12.1	10.9	8.0	5.8	4.2	2.9	2.0	1.3	A335	5 Cr–$\frac{1}{2}$ Mo-Ti
17.4	17.2	17.1	16.8	16.6	16.3	13.2	12.8	12.1	10.9	8.0	5.8	4.2	2.9	2.0	1.3	A335	7 Cr–$\frac{1}{2}$ Mo
17.4	17.2	17.1	16.8	16.6	16.3	13.2	12.8	12.1	11.4	10.6	7.4	5.0	3.3	2.2	1.5	A335	9 Cr–1 Mo
18.0	17.5	17.2	16.7	16.2	15.6	15.0	15.0	14.4	13.1	11.0	7.8	5.5	4.0	2.5	1.2	A335	$1\frac{1}{4}$ Cr–$\frac{1}{2}$ Mo
18.0	17.5	17.2	16.7	16.2	15.6	15.0	14.8	14.2	13.1	11.0	7.5	5.0	2.8	1.6	1.0	A335	1 Cr–$\frac{1}{2}$ Mo
17.6	17.0	16.5	15.9	15.6	15.3	15.0	14.4	13.8	12.5	10.0	6.3	4.0	2.4	–	–	A335	$1\frac{1}{2}$ Si–$\frac{1}{2}$ Mo
18.0	17.5	17.2	16.7	16.2	15.6	15.2	15.0	14.0	12.0	9.0	7.0	5.5	4.0	2.7	1.5	A335	3 Cr–1 Mo

Material	Spec																
$2\frac{1}{4}$Cr-1Mo	A335	2.0	3.0	4.2	5.8	7.8	11.0	12.8	14.5	15.2	17.9	17.9	17.9	17.9	17.9	17.9	18.0
C-$\frac{1}{2}$Mo	A369	–	–	2.4	4.0	4.8	9.2	12.7	13.1	13.5	13.8	15.1	15.4	15.7	16.3	16.9	17.5
$\frac{3}{4}$Cr-$\frac{1}{2}$Mo	A369	–	–	2.4	4.0	5.9	9.2	12.8	13.1	13.5	13.8	15.1	15.4	15.7	16.3	16.9	17.5
2Cr-$\frac{1}{2}$Mo	A369	1.0	1.4	2.6	4.2	6.2	10.0	12.5	13.1	13.5	13.9	15.1	15.4	15.7	16.3	16.4	17.5
5Cr-$\frac{1}{2}$Mo	A369	1.3	2.0	2.9	4.2	5.8	8.0	10.9	12.1	12.8	13.2	16.3	16.6	16.8	17.1	17.2	17.4
7Cr-$\frac{1}{2}$Mo	A369	1.3	2.0	2.9	4.2	5.8	8.0	10.9	12.1	12.8	13.2	16.3	16.6	16.8	17.1	17.2	17.4
9Cr-1Mo	A369	1.5	2.2	3.3	5.0	7.4	10.0	11.4	12.1	12.8	13.2	16.3	16.6	16.8	17.1	17.2	17.4
$1\frac{1}{4}$Cr-$\frac{1}{2}$Mo	A369	1.2	2.5	4.0	5.5	7.8	11.0	12.8	14.5	15.0	15.2	15.6	16.2	16.7	17.2	17.5	18.0
1Cr-$\frac{1}{2}$Mo	A369	1.0	1.6	2.8	5.0	7.5	11.0	12.8	14.5	15.0	15.2	15.6	16.2	16.7	17.2	17.5	18.0
3Cr-1Mo	A369	1.5	2.7	4.0	5.5	7.0	9.0	12.0	14.0	15.0	15.2	15.6	16.2	16.7	17.2	17.5	18.0
$2\frac{1}{4}$Cr-1Mo	A369	1.0	3.0	4.2	5.8	7.8	11.0	12.8	14.5	15.2	17.9	17.9	17.9	17.9	17.9	17.9	18.0
Centrifugally Cast Pipe																	
C-$\frac{1}{2}$Mo	A426	–	–	2.4	4.0	6.3	10.0	12.5	14.4	15.7	16.3	20.0	20.4	20.7	21.3	21.7	21.7
$\frac{3}{4}$Cr-$\frac{1}{2}$Mo	A426	–	–	2.4	4.0	6.3	10.0	12.5	13.2	13.5	13.9	14.2	14.6	14.9	15.6	16.3	17.0
5Cr-$\frac{1}{2}$Mo	A426	1.3	2.0	3.1	4.2	5.6	7.6	10.4	12.8	14.5	16.0	17.5	19.0	20.1	22.1	24.1	26.1
5Cr-$\frac{1}{2}$Mo-$1\frac{1}{2}$Si	A426	1.2	1.8	2.5	3.5	5.5	9.0	10.9	12.4	12.8	13.3	13.7	14.1	14.5	15.4	16.2	17.1
7Cr-$\frac{1}{2}$Mo	A426	1.2	1.8	2.5	3.5	5.0	7.0	9.5	11.5	12.5	13.1	13.6	14.0	14.5	15.3	16.2	17.0
9Cr-1Mo	A426	1.5	2.2	3.3	5.5	8.5	10.7	15.0	17.3	19.4	21.0	22.0	22.5	22.5	22.5	22.5	22.5
$1\frac{1}{4}$Cr-$\frac{1}{2}$Mo	A426	1.2	2.5	4.0	5.5	7.8	11.0	13.1	14.4	15.0	15.5	20.9	21.6	22.3	22.9	23.3	23.3
1Cr-$\frac{1}{2}$Mo	A426	1.0	1.6	2.8	5.0	7.5	11.0	13.1	14.2	14.8	15.1	15.4	15.7	15.9	16.5	17.1	17.6
$1\frac{1}{2}$Si-$\frac{1}{2}$Mo	A426	–	–	4.2	4.0	6.3	10.0	12.5	13.8	14.4	15.0	15.3	15.6	15.9	16.5	17.0	17.6
$12\frac{3}{4}$Cr	A426	–	–	–	–	–	–	–	–	–	–	–	–	–	–	–	–
3Cr-1Mo	A426	1.5	2.7	4.0	5.5	7.0	9.0	12.0	13.2	13.9	14.5	14.8	15.2	15.5	16.1	16.8	17.4
$2\frac{1}{4}$Cr-1Mo	A426	2.0	3.0	4.2	5.8	7.8	11.0	14.0	16.0	17.5	17.5	20.9	21.6	23.3	22.9	23.3	23.3

TABLE A3 (Continued)
Allowable Stresses in Tension for Metals, SE, KSI

Material	Speci-fication	P-No. (37)	Grade	Min. Tensile Str. (ksi)	Min. Yield Str. (ksi)	Notes	Min. Temp. (26)	Min. Temp. to 100	200	300	400	500	600
Stainless Steel													
Seamless Pipe and Tubes													
26 Cr–3 Ni–1 Mo Tubes	A268	10E	TP329	90.0	70.0	28	–20	30.0	–	–	–	–	–
12 Cr–Al Tubes	A268	7	TP405	60.0	30.0	28	–20	20.0	18.4	17.7	17.4	17.2	16.8
11 Cr–Ti Tubes	A268	6	TP409	60.0	30.0	28	–20	20.0	–	–	–	–	–
13 Cr Tubes	A268	6	TP410	60.0	30.0	28	–20	20.0	18.4	17.7	17.4	17.2	16.8
16 Cr Tubes	A268	7	TP430	60.0	35.0	28, 38	–20	20.0	20.0	19.6	19.2	19.0	18.5
18 Cr–Ti Tubes	A268	7	TP430Ti	60.0	40.0	28, 38	–20	20.0	–	–	–	–	–
20 Cr–Cu Tubes	A268	10	TP443	70.0	40.0	28	–20	23.3	23.3	21.4	20.4	19.4	18.4
27 Cr Tubes	A268	10E	TP446	70.0	40.0	28	–20	23.3	23.3	21.4	20.4	19.4	18.4
18 Cr–8 Ni Tubes	A269	8	TP304	75.0	30.0	6b, 20, 23, 43	–425	20.0	20.0	20.0	18.7	17.5	16.4
18 Cr–8 Ni Tubes	A269	8	TP304L	70.0	25.0	43	–425	16.7	16.7	16.7	15.8	14.8	14.0
16 Cr–12 Ni–2 Mo Tubes	A269	8	TP316	75.0	30.0	20, 43	–325	20.0	20.0	20.0	19.3	17.9	17.0
16 Cr–12 Ni–2 Mo Tubes	A269	8	TP316L	70.0	25.0	43	–325	16.7	16.7	16.7	15.5	14.4	13.5
18 Cr–8 Ni Pipe	A312	8	TP304	75.0	30.0	6a, 20, 23, 30	–425	20.0	20.0	20.0	18.7	17.5	16.4
18 Cr–8 Ni Pipe	A312	8	TP304H	75.0	30.0	23	–325	20.0	20.0	20.0	18.7	17.5	16.4
18 Cr–8 Ni Pipe	A312	8	TP304L	70.0	25.0	–	–425	16.7	16.7	16.7	15.8	14.8	14.0
23 Cr–12 Ni Pipe	A312	8	TP309	75.0	30.0	28, 41, 52	–325|	20.0	20.0	20.0	20.0	20.0	19.2

Material	Spec	No.	Grade			Notes							
25 Cr–20 Ni Pipe	A312	8	TP310	75.0	30.0	28, 41, 52	–325	20.0	20.0	20.0	20.0	20.0	19.2
25 Cr–20 Ni Pipe	A312	8	TP310	75.0	30.0	6, 28, 41, 52	–325	20.0	20.0	20.0	20.0	20.0	19.2
16 Cr–12 Ni–2 Mo Pipe	A312	8	TP316	75.0	30.0	20, 23	–325	20.0	20.0	20.0	19.3	17.9	17.0
16 Cr–12 Ni–2 Mo Pipe	A312	8	TP316H	75.0	30.0	23	–325	20.0	20.0	20.0	19.3	17.9	17.0
16 Cr–12 Ni–2 Mo Pipe	A312	8	TP316L	70.0	25.0	–	–325	16.7	16.7	16.7	15.5	14.4	13.5
18 Cr–13 Ni–3 Mo	A312	8	TP317	75.0	30.0	20, 23	–325	20.0	20.0	20.0	19.3	17.9	17.0
18 Cr–10 Ni–Ti Pipe	A313	8	TP321	75.0	30.0	6a, 20	–325	20.0	20.0	20.0	18.6	17.3	16.4
18 Cr–10 Ni–Ti Pipe	A312	8	TP321H	75.0	30.0	–	–325	20.0	20.0	20.0	18.6	17.3	16.4
18 Cr–10 Ni–Cb	A312	8	TP347	75.0	30.0	6a, 20	–425	20.0	20.0	20.0	20.0	19.9	19.3
18 Cr–10 Ni–Cb	A312	8	TP347H	75.0	30.0	–	–325	20.0	20.0	20.0	20.0	19.9	19.3
18 Cr–10 Ni–Cb Pipe	A312	8	TP348	75.0	30.0	6a, 20	–325	20.0	20.0	20.0	20.0	19.9	19.3
18 Cr–10 Ni–Cb Pipe	A312	9	TP348H	75.0	30.0	–	–325	20.0	20.0	20.0	20.0	19.9	19.3
18 Cr–8 Ni Pipe	A376	8	TP304	75.0	30.0	6b, 20, 23, 30, 36	–425	20.0	20.0	20.0	18.7	17.5	16.4
18 Cr–8 Ni Pipe	A376	8	TP304H	75.0	30.0	23	–325	20.0	20.0	20.0	18.7	17.5	16.4
16 Cr–12 Ni–2 Mo Pipe	A376	8	TP316	75.0	30.0	6b, 20, 23, 36	–325	20.0	20.0	20.0	19.3	17.9	17.0
16 Cr–12 Ni–2 Mo Pipe	A376	8	TP316H	75.0	30.0	23	–325	20.0	20.0	20.0	19.3	17.9	17.0
18 Cr–10 Ni–Ti Pipe	A376	8	TP321	75.0	30.0	6a, 20, 36	–325	20.0	20.0	20.0	18.6	17.3	16.4

TABLE A3 (Continued)
Allowable Stresses in Tension for Metals, SE, KSI

650	700	750	800	850	900	950	1000	1050	1100	1150	1200	1250	1300	1350	1400	1450	1500	Speci-fication	Material
																			Stainless Steel Seamless Pipe and Tubes
–	–	–	–	–	–	–	–	–	–	–	–	–	–	–	–	–	–	A268	26 Cr–3 Ni–1 Mo Tubes
16.5	16.2	11.6	11.1	10.4	9.6	8.4	4.0	–	–	–	–	–	–	–	–	–	–	A268	12 Cr–Al Tubes
–	–	–	–	–	–	–	–	–	–	–	–	–	–	–	–	–	–	A268	11 Cr–Tl Tubes
16.5	16.2	11.6	11.1	10.4	9.6	8.4	6.4	4.4	2.9	1.8	1.0	–	–	–	–	–	–	A268	13 Cr Tubes
18.2	17.6	11.6	11.1	10.4	9.6	8.5	6.5	4.5	3.2	2.4	1.7	–	–	–	–	–	–	A268	16 Cr Tubes
–	–	–	–	–	–	–	–	–	–	–	–	–	–	–	–	–	–	A268	18 Cr–Ti Tubes
18.0	17.5	16.9	16.2	15.1	13.0	6.8	4.5	–	–	–	–	–	–	–	–	–	–	A268	20 Cr–Cu Tubes
18.0	17.5	16.9	16.2	15.1	13.0	6.8	4.5	–	–	–	–	–	–	–	–	–	–	A268	27 Cr Tubes
16.2	16.0	15.6	15.2	14.9	14.6	14.4	13.8	12.2	9.7	7.7	6.0	4.7	3.7	2.9	2.3	1.8	1.4	A269	18 Cr–8 Ni Tubes
13.7	13.5	13.3	13.0	12.8	11.9	9.9	7.8	6.3	5.1	4.0	3.2	2.6	2.1	1.7	1.1	1.0	0.9	A269	18 Cr–8 Ni Tubes
16.7	16.3	16.1	15.9	15.7	15.5	15.4	15.3	14.5	12.4	9.8	7.4	5.5	4.1	3.1	2.3	1.7	1.3	A269	16 Cr–12 Ni–2 Mo Tubes
13.2	12.9	12.6	12.4	12.1	11.8	11.5	11.2	10.8	10.2	8.8	6.4	4.7	3.5	2.5	1.8	1.3	1.0	A269	16 Cr–12 Ni–2 Mo Tubes
16.2	16.0	15.6	15.2	14.9	14.6	14.6	13.8	12.2	9.7	7.7	6.0	4.7	3.7	2.9	2.3	1.8	1.4	A312	18 Cr–8 Ni Pipe
16.2	16.0	15.6	15.2	14.9	14.6	14.6	13.8	12.2	9.7	7.7	6.0	4.7	3.7	2.9	2.3	1.8	1.4	A312	18 Cr–8 Ni Pipe
13.7	13.5	13.3	13.0	12.8	11.9	9.9	7.8	6.3	5.1	4.0	3.2	2.6	2.1	1.7	1.1	1.0	0.9	A312	18 Cr–8 Ni Tubes
18.8	18.3	18.0	17.5	14.6	13.9	12.5	10.5	8.5	6.5	5.0	3.8	2.9	2.3	1.8	1.3	0.9	0.7	A312	23 Cr–12 Ni Pipe
18.8	18.3	18.0	17.5	14.6	13.9	12.5	11.0	7.1	5.0	3.6	2.5	1.5	0.8	0.5	0.4	0.3	0.2	A312	25 Cr–20 Ni Pipe
18.8	18.3	18.0	17.5	14.6	13.9	12.5	11.0	9.8	8.5	7.3	6.0	4.8	3.5	2.3	1.6	1.1	0.8	A312	25 Cr–20 Ni Pipe

V1	V2	V3	V4	V5	V6	V7	V8	V9	V10	V11	V12	V13	V14	V15	V16	V17	V18	Spec	Material
16.7	16.3	16.1	15.9	15.7	15.5	15.4	15.3	14.5	12.4	9.8	7.4	5.5	4.1	3.1	2.3	1.7	1.3	A312	16 Cr–12 Ni–2 Mo Pipe
16.7	16.3	16.1	15.9	15.7	15.5	15.4	15.3	14.5	12.4	9.8	7.4	5.5	4.1	3.1	2.3	1.7	1.3	A312	16 Cr–12 Ni–2 Mo Pipe
13.2	12.9	12.6	12.4	12.1	11.8	11.5	11.2	10.8	10.2	8.8	6.4	4.7	3.5	2.5	1.8	1.3	1.0	A312	16 Cr–12 Ni–2 Mo Pipe
16.7	16.3	16.1	15.9	15.7	15.5	15.4	15.3	14.5	12.4	9.8	7.4	5.5	4.1	3.2	2.3	1.7	1.3	A312	18 Cr–13 Ni–3 Mo Pipe
16.1	15.8	15.7	15.5	15.3	15.2	15.1	13.8|	9.6	6.9	5.0	3.6	2.6	1.7	1.1	0.8	0.5	0.3	A312	18 Cr–10 Ni–Ti Pipe
16.1	15.8	15.7	15.5	15.3	15.2	15.1	14.0	11.7	9.0	6.9	5.4	4.1	3.2	2.5	1.9	1.5	1.1	A312	18 Cr–10 Ni–Ti Pipe
19.0	18.6	18.5	18.3	15.4	14.9	14.8	14.0|	12.1	9.1	6.1	4.4	3.3	2.2	1.5	1.2	0.9	0.8	A312	18 Cr–10 Ni–Cb Pipe
19.0	18.6	18.5	18.3	18.2	18.1	18.1	18.0	17.1	14.2	10.5	7.9	5.9	4.4	3.2	2.5	1.8	1.3	A312	18 Cr–10 Ni–Cb Pipe
19.0	18.6	18.5	18.3	15.4	14.9	14.8	14.0|	12.1	9.1	6.1	4.4	3.3	2.2	1.5	1.2	0.9	0.8	A312	18 Cr–10 Ni–Cb Pipe
19.0	18.6	18.5	18.3	18.2	18.1	18.1	18.0	17.1	14.2	10.5	7.9	5.9	4.4	3.2	2.5	1.8	1.3	A312	18 Cr–10 Ni–Cb Pipe
16.2	16.0	15.6	15.2	14.9	14.6	14.4	13.8|	12.2	9.7	7.7	6.0	4.7	3.7	2.9	2.3	1.8	1.4	A376	18 Cr–8 Ni Pipe
16.2	16.0	15.6	15.2	14.9	14.6	14.4	13.8	12.2	9.7	7.7	6.0	4.7	3.7	2.9	2.3	1.8	1.4	A376	18 Cr–8 Ni Pipe
16.7	16.3	16.1	15.9	15.7	15.5	15.4	15.3|	14.5	12.4	9.8	7.4	5.5	4.1	3.1	2.3	1.7	1.3	A376	16 Cr–12 Ni–2 Mo Pipe
16.7	16.3	16.1	15.9	15.7	15.5	15.4	15.3	14.5	12.4	9.8	7.4	5.5	4.1	3.1	2.3	1.7	1.3	A376	16 Cr–12 Ni–2 Mo Pipe
16.1	15.8	15.7	15.5	15.3	15.2	15.1	13.8|	9.6	6.9	5.0	3.6	2.3	1.7	1.1	0.8	0.5	0.3	A376	18 Cr–10 Ni–Ti Pipe

TABLE A3 (Continued)
Allowable Stresses in Tension for Metals, SE, KSI

Stainless Steel (Cont.)
Seamless Pipe and Tubes (Cont.)

Material	Speci-fication	P-No. (37)	Grade	Min. Tensile Str. (ksi)	Min. Yield Str. (ksi)	Notes	Min. Temp. (26)	Min. Temp. to 100	200	300	400	500	600
18 Cr–10 Ni–Ti Pipe	A376	8	TP321H	75.0	30.0	–	–325	20.0	20.0	20.0	18.6	17.3	16.4
18 Cr–10 Ni–Cb Pipe	A376	8	TP347	75.0	30.0	6a, 20, 36	–425|	20.0	20.0	20.0	20.0	19.9	19.3
18 Cr–10 Ni–Cb Pipe	A376	8	TP347H	75.0	30.0	–	–325|	20.0	20.0	20.0	20.0	19.9	19.3
18 Cr–10 Ni–Cb Pipe	A376	8	TP348	75.0	30.0	6a, 20, 36	–325|	20.0	20.0	20.0	20.0	19.9	19.3
16 Cr–8 Ni–2 Mo Pipe	A376	8	16–8–2H	75.0	30.0	6b, 12, 23	–325	20.0	–	–	–	–	–
18 Cr–8 Ni Pipe	A430	8	FP304	70.0	30.0	6b, 23, 36	–425|	20.0	20.0	20.0	18.7	17.5	16.4
18 Cr–8 Ni Pipe	A430	8	FP304H	70.0	30.0	23, 36	–325|	20.0	20.0	20.0	18.7	17.5	16.4
16 Cr–12 Ni–2 Mo Pipe	A430	8	FP316	70.0	30.0	6b, 23, 36	–325|	20.0	20.0	20.0	19.3	17.9	17.0
16 Cr–12 Ni–2 Mo Pipe	A430	8	FP316H	70.0	30.0	23, 36	–325|	20.0	20.0	20.0	19.3	17.9	17.0
18 Cr–10 Ni–Ti Pipe	A430	8	FP321	70.0	30.0	6a, 36	–325|	20.0	20.0	20.0	18.6	17.3	16.4
18 Cr–10 Ni–Ti Pipe	A430	8	FP321H	70.0	30.0	36	–325|	20.0	20.0	20.0	18.6	17.3	16.4
18 Cr–10 Ni–Cb Pipe	A430	8	FP347	70.0	30.0	6a, 36	–425|	20.0	20.0	20.0	19.2	18.6	18.3
18 Cr–10 Ni–Cb Pipe	A430	8	FP347H	70.0	30.0	36	–325|	20.0	20.0	20.0	19.2	18.6	18.3

650	700	750	800	850	900	950	1000	1050	1100	1150	1200	1250	1300	1350	1400	1450	1500	Specification	Material
																			Stainless Steel (Cont.)
																		Seamless Pipe and Tubes (Cont.)	
16.1	15.8	15.7	15.5	15.3	15.2	15.1	14.0	11.7	9.0	6.9	5.4	4.1	3.2	2.5	1.9	1.5	1.1	A376	18 Cr–10 Ni–Ti Pipe
19.0	18.6	18.5	18.3	15.4	14.9	14.8	14.0	12.1	9.1	6.1	4.4	3.3	2.2	1.5	1.2	0.9	0.8	A376	18 Cr–10 Ni–Cb Pipe
19.0	18.6	18.5	18.3	18.2	18.1	18.1	18.0	17.1	14.2	10.5	7.9	5.9	4.4	3.2	2.5	1.8	1.3	A376	18 Cr–10 Ni–Cb Pipe
19.0	18.6	18.5	18.3	15.4	14.9	14.8	14.0	12.1	9.1	6.1	4.4	3.3	2.2	1.5	1.2	0.9	0.8	A376	18 Cr–10 Ni–Cb Pipe
–	–	–	–	–	–	–	–	–	–	–	–	–	–	–	–	–	–	A376	16 Cr–8 Ni–2 Mo Pipe
16.2	16.0	15.6	15.2	14.9	14.6	14.4	13.8	12.2	9.7	7.7	6.0	4.7	3.7	2.9	2.3	1.8	1.4	A430	18 Cr–8 Ni Pipe
16.2	16.0	15.6	15.2	14.9	14.6	14.4	13.8	12.2	9.7	7.7	6.0	4.7	3.7	2.9	2.3	1.8	1.4	A430	18 Cr–8 Ni Pipe
16.7	16.3	16.1	15.9	15.7	15.5	15.4	15.3	14.5	12.4	9.8	7.4	5.5	4.1	3.1	2.3	1.7	1.3	A430	16 Cr–12 Ni–2 Mo Pipe
16.7	16.3	16.1	15.9	15.7	15.5	15.4	15.3	14.5	12.4	9.8	7.4	5.5	4.1	3.1	2.3	1.7	1.3	A430	16 Cr–12 Ni–2 Mo Pipe
16.1	15.8	15.7	15.5	15.3	15.2	15.1	13.8	9.6	6.9	5.0	3.6	2.6	1.7	1.1	0.8	0.5	0.3	A430	18 Cr–10 Ni–Ti Pipe
16.1	15.8	15.7	15.5	15.3	15.2	15.1	14.0	11.7	9.0	6.9	5.4	4.1	3.2	2.5	1.9	1.5	1.1	A430	18 Cr–10 Ni–Ti Pipe
18.2	18.2	18.2	18.2	14.0	13.9	13.8	13.2	12.1	9.1	6.1	4.4	3.3	2.2	1.5	1.2	0.9	0.8	A430	18 Cr–10 Ni–Cb Pipe
18.2	18.2	18.2	18.2	18.1	18.1	18.1	18.0	17.1	14.3	10.5	7.9	5.9	4.4	3.2	2.5	1.8	1.3	A430	18 Cr–10 Ni–Cb Pipe

TABLE A4 Properties and Weights of Pipe

Nominal Size Outside Diameter inches D	Weight Designation and/or Schedule Number	Average Wall Thickness inches t	Minimum Wall Thickness $(=\frac{7}{8}t)$ inches t_m	Inside Diameter inches d	Cross-Sectional Metal Area sq. inches A	Moment of Inertia inches^4 I	Section Modulus inches^3 Z	Bend Characteristic per Unit Bend Radius 1/ft h/R	Radius of Gyration inches r_g	Weight of Pipe w_p lb per ft	Weight of Water w_w lb per ft
$\frac{1}{8}''$ 0.405	10S	0.049	0.043	0.307	0.055	0.0009	0.0043	18.6	0.127	0.186	0.032
	Std. 40 40S	0.068	0.030	0.269	0.072	0.0011	0.0052	28.7	0.122	0.245	0.025
	XS 80 80S	0.095	0.083	0.215	0.092	0.0012	0.0060	47.5	0.115	0.315	0.016
$\frac{1}{4}''$ 0.540	10S	0.065	0.057	0.410	0.097	0.0028	0.0103	13.8	0.169	0.330	0.057
	Std. 40 40S	0.088	0.077	0.364	0.125	0.0033	0.0123	20.7	0.163	0.425	0.045
	XS 80 80S	0.119	0.104	0.302	0.157	0.0038	0.0140	32.2	0.155	0.535	0.031
$\frac{3}{8}''$ 0.675	10S	0.065	0.057	0.545	0.124	0.0059	0.0174	8.38	0.217	0.423	0.101
	Std. 40 40S	0.091	0.080	0.493	0.167	0.0073	0.0216	12.81	0.209	0.568	0.083
	XS 80 80S	0.126	0.110	0.423	0.217	0.0086	0.0255	20.1	0.199	0.739	0.061
$\frac{1}{2}''$ 0.840	10S	0.083	0.073	0.674	0.197	0.0143	0.0341	6.95	0.269	0.671	0.154
	Std. 40 40S	0.109	0.095	0.622	0.250	0.0171	0.0407	9.79	0.261	0.851	0.132
	XS 80 80S	0.147	0.129	0.546	0.320	0.0201	0.0478	14.7	0.250	1.09	0.101
	160	0.187	0.164	0.466	0.384	0.0221	0.0527	21.1	0.240	1.30	0.074
	XXS	0.294	0.258	0.252	0.504	0.0243	0.0577	47.3	0.219	1.72	0.022
$\frac{3}{4}''$ 1.050	5S	0.065	0.057	0.920	0.201	0.0245	0.0467	3.22	0.349	0.684	0.288
	10S	0.083	0.073	0.884	0.252	0.0297	0.0566	4.26	0.343	0.857	0.266
	Std. 40 40S	0.113	0.099	0.824	0.333	0.0370	0.0706	6.18	0.334	1.13	0.231

Size	Sched.	S										
	XS 80	80S	0.154	0.135	0.742	0.434	0.0448	0.0853	9.21	0.321	1.47	0.187
	160		0.218	0.191	0.614	0.570	0.0527	0.100	15.1	0.304	1.94	0.128
	XXS		0.308	0.270	0.431	0.718	0.0579	0.110	26.9	0.284	2.44	0.064
1" 1.315		5S	0.065	0.057	1.185	0.255	0.0500	0.076	2.00	0.443	0.868	0.478
		10S	0.109	0.095	1.097	0.413	0.0757	0.115	3.60	0.428	1.40	0.409
	Std. 40	40S	0.133	0.116	1.049	0.494	0.0874	0.133	4.57	0.420	1.68	0.374
	XS 80	80S	0.179	0.157	0.957	0.639	0.106	0.161	6.66	0.407	2.17	0.311
	160		0.250	0.219	0.815	0.836	0.125	0.190	10.58	0.387	2.84	0.226
	XXS		0.358	0.313	0.599	1.08	0.141	0.214	18.76	0.361	3.66	0.122
1¼ 1.660		5S	0.065	0.057	1.530	0.33	0.104	0.125	1.23	0.56	1.11	0.80
		10S	0.109	0.095	1.442	0.53	0.161	0.193	2.17	0.55	1.81	0.71
	Std. 40	40S	0.140	0.123	1.380	0.67	0.195	0.235	2.91	0.54	2.27	0.65
	XS 80	80S	0.191	0.167	1.278	0.88	0.242	0.291	4.25	0.52	3.00	0.56
	160		0.250	0.219	1.160	1.11	0.284	0.342	6.04	0.51	3.76	0.46
	XXS		0.382	0.334	0.896	1.53	0.341	0.411	11.2	0.47	5.22	0.27
1½ 1.900		5S	0.065	0.057	1.770	0.38	0.158	0.166	0.927	0.65	1.27	1.07
		10S	0.109	0.095	1.682	0.61	0.247	0.260	1.63	0.63	2.09	0.96
	Std. 40	40S	0.145	0.127	1.610	0.80	0.310	0.326	2.26	0.62	2.72	0.88
	XS 80	80S	0.200	0.175	1.500	1.07	0.391	0.412	3.32	0.61	3.63	0.77
	160		0.281	0.246	1.338	1.43	0.483	0.508	5.15	0.58	4.87	0.61
	XXS		0.400	0.350	1.100	1.89	0.568	0.598	8.53	0.55	6.41	0.41
		5S	0.065	0.057	2.245	0.47	0.315	0.265	0.585	0.82	1.60	1.72
		10S	0.109	0.095	2.157	0.78	0.499	0.420	1.02	0.80	2.64	1.58
	Std. 40	40S	0.154	0.135	2.067	1.07	0.666	0.561	1.50	0.79	3.65	1.45

TABLE A4 (Continued)
Properties and Weights of Pipe

Nominal Size Outside Diameter inches D	Weight Designation and/or Schedule Number			Average Wall Thickness inches t	Minimum Wall Thickness ($=\frac{7}{8}t$) inches t_m	Inside Diameter inches d	Cross-Sectional Metal Area sq. inches A	Moment of Inertia inches^4 I	Section Modulus inches^3 Z	Bend Characteristic per Unit Bend Radius 1/ft h/R	Radius of Gyration inches r_g	Weight of Pipe w_p lb per ft	Weight of Water w_w lb per ft
2" 2.375	XS	80	80S	0.218	0.191	1.939	1.48	0.868	0.731	2.25	0.77	5.02	1.28
		160		0.343	0.300	1.689	2.19	1.16	0.979	3.99	0.73	7.45	0.97
	XXS			0.436	0.382	1.503	2.66	1.31	1.10	5.57	0.70	9.03	0.77
$2\frac{1}{2}''$ 2.875			5S	0.083	0.073	2.709	0.73	0.710	0.494	0.511	0.99	2.48	2.50
			10S	0.120	0.105	2.635	1.04	0.988	0.687	0.759	0.98	3.53	2.36
	Std.	40	40S	0.203	0.178	2.469	1.70	1.53	1.03	1.37	0.95	5.79	2.08
	XS	80	80S	0.276	0.242	2.323	2.25	1.93	1.34	1.96	0.92	7.66	1.84
		160		0.375	0.328	2.125	2.95	2.35	1.64	2.88	0.89	10.0	1.54
	XXS			0.552	0.483	1.771	4.03	2.87	2.00	4.91	0.84	13.7	1.07
3.500			5S	0.083	0.073	3.334	0.89	1.30	0.744	0.341	1.21	3.03	3.78
			10S	0.120	0.105	3.260	1.27	1.82	1.04	0.504	1.20	4.33	3.61
	Std.	40	40S	0.216	0.189	3.068	2.23	3.02	1.72	0.961	1.16	7.58	3.20
	XS	80	80S	0.300	0.263	2.900	3.02	3.90	2.23	1.41	1.14	10.3	2.86
		160		0.438	0.382	2.624	4.21	5.04	2.88	2.24	1.09	14.3	2.34

Nominal Size	OD	Schedule										
		XXS	0.600	0.525	2.300	5.47	5.99	3.43	3.42	1.05	18.6	1.80
3½"	4.000	5S	0.83	0.073	3.834	1.02	1.96	0.980	0.260	1.39	3.47	5.00
		10S	0.120	0.195	3.760	1.46	2.76	1.38	0.383	1.37	4.97	4.81
		40 Std.	0.226	0.198	3.548	2.68	4.79	2.39	0.762	1.34	9.11	4.28
		80 XS	0.318	0.278	3.364	3.68	6.28	3.14	1.13	1.31	12.5	3.85
		XXS	0.636	0.557	2.728	6.72	9.85	4.93	2.70	1.21	22.9	2.53
4"	4.500	5S	0.083	0.073	4.334	1.15	2.81	1.25	0.204	1.56	3.92	6.40
		10S	0.120	0.105	4.260	1.65	3.96	1.76	0.300	1.55	5.61	6.17
		40 Std.	0.237	0.207	4.026	3.17	7.23	3.21	0.626	1.51	10.8	5.51
		80 XS	0.337	0.295	3.826	4.41	9.61	4.27	0.933	1.48	15.0	4.98
		120	0.438	0.382	3.624	5.59	11.7	5.18	1.27	1.45	19.0	4.47
		160	0.531	0.465	3.438	6.62	13.3	5.90	1.62	1.42	22.5	4.02
		XXS	0.674	0.590	3.152	8.10	15.3	6.79	2.21	1.37	27.5	3.38
5"	5.563	5S	0.109	0.095	5.345	1.87	6.95	2.50	0.176	1.93	6.35	9.73
		10S	0.134	0.117	5.295	2.29	8.43	3.03	0.218	1.92	7.77	9.53
		40 Std.	0.258	0.226	5.047	4.30	15.2	5.45	0.440	1.88	14.6	8.66
		80 XS	0.375	0.328	4.813	6.11	20.7	7.43	0.669	1.84	20.8	7.88
		120	0.500	0.438	4.563	7.95	25.7	9.25	0.936	1.80	27.0	7.09
		160	0.625	0.547	4.313	9.70	30.0	10.8	1.23	1.76	33.0	6.33
		XXS	0.750	0.655	4.063	11.3	33.6	12.1	1.55	1.72	38.6	5.62
6"	6.625	5S	0.109	0.095	6.407	2.23	11.9	3.58	0.123	2.30	5.37	14.0
		10S	0.134	0.117	6.357	2.73	14.4	4.35	0.153	2.30	9.29	13.7
		40 Std.	0.280	0.245	6.065	5.58	28.1	8.50	0.334	2.25	19.0	12.5

TABLE A4 (Continued)
Properties and Weights of Pipe

Nominal Size Outside Diameter inches D	Weight Designation and/or Schedule Number	Average Wall Thickness inches t	Minimum Wall Thickness $(=\frac{7}{8}t)$ inches t_m	Inside Diameter inches d	Cross-Sectional Metal Area sq. inches A	Moment of Inertia inches⁴ I	Section Modulus inches³ Z	Bend Characteristic per Unit Bend Radius 1/ft h/R	Radius of Gyration inches r_g	Weight of Pipe w_p lb per ft	Weight of Water w_w lb per ft
	XS 80 80S	0.432	0.378	5.761	8.40	40.5	12.2	0.541	2.20	28.6	11.3
	120	0.562	0.492	5.501	10.7	49.6	15.0	0.735	2.15	36.4	10.3
	160	0.718	0.628	5.189	13.3	59.0	17.8	9.988	2.10	45.3	9.16
	XXS	0.864	0.756	4.897	15.6	66.3	20.0	1.25	2.06	53.2	8.14
	5S	0.109	0.095	8.407	2.92	26.5	6.13	0.073	3.01	9.91	24.1
	10S	0.148	0.130	8.329	3.94	35.4	8.21	0.099	3.00	13.4	23.6
8″	20	0.250	0.219	8.125	6.58	57.7	13.4	0.171	2.96	22.4	22.5
	30	0.277	0.242	8.071	7.26	63.4	14.7	0.191	2.95	24.7	22.2
8.625	Std. 40 40S	0.322	0.282	7.981	8.40	72.5	16.8	0.224	2.94	28.6	21.7
	60	0.406	0.355	7.813	10.5	88.8	20.6	0.289	2.91	35.6	20.8
	XS 80 80S	0.500	0.438	7.625	12.8	106	24.5	0.364	2.88	43.4	19.8
	100	0.593	0.519	7.439	15.0	121	28.1	0.441	2.85	50.9	18.8
	120	0.718	0.628	7.189	17.8	141	32.6	0.551	2.81	60.6	17.6
	140	0.812	0.711	7.001	19.9	154	35.6	0.639	2.78	67.8	16.7

NPS (OD)	Sched.	t		ID							
	XXS	0.875	0.766	6.875	21.3	162	37.6	0.699	2.76	72.4	16.1
	160	0.906	0.793	6.813	22.0	166	38.5	0.730	2.75	74.7	15.8
10″ 10.750	5S	0.134	0.117	10.483	4.52	63.7	11.9	0.057	3.75	15.2	37.4
	10S	0.165	0.144	10.420	5.49	76.9	14.3	0.071	3.74	18.7	36.9
	20	0.250	0.219	10.250	8.26	114	21.2	0.109	3.71	28.0	35.7
	30	0.307	0.269	10.136	10.1	138	25.6	0.135	3.69	34.2	34.9
	40, Std., 40S	0.365	0.319	10.020	11.9	161	29.9	0.163	3.67	40.5	34.1
	60, XS, 80S	0.500	0.438	9.750	16.1	212	39.4	0.228	3.63	54.7	32.3
	80	0.593	0.519	9.564	18.9	245	45.5	0.276	3.60	64.3	31.1
		0.625	0.547	9.500	19.9	256	47.6	0.293	3.59	67.5	30.7
	100	0.718	0.628	9.314	22.6	286	53.2	0.342	3.56	76.9	29.5
		0.750	0.655	9.250	23.6	296	55.1	0.360	3.55	80.1	29.1
	120	0.843	0.738	9.064	26.2	324	60.3	0.412	3.52	80.2	27.9
		0.875	0.766	9.000	27.1	333	62.0	0.431	3.51	92.3	27.5
	140	1.000	0.875	8.750	30.6	368	68.4	0.505	3.47	104	26.0
	160	1.125	0.984	8.500	34.0	399	74.3	0.583	3.43	116	24.6
12″ 12.750	5S	0.165	0.144	12.420	6.52	129	20.3	0.050	4.45	19.6	52.5
	10S	0.180	0.158	12.390	7.11	141	22.0	0.055	4.44	24.2	52.2
	20	0.250	0.219	12.250	9.82	192	30.0	0.077	4.42	33.4	51.1
	30	0.330	0.289	12.090	12.9	249	39.0	0.103	4.39	43.8	49.7
	Std., 40S	0.375	0.328	12.000	14.6	279	43.8	0.118	4.38	49.6	49.0
	40	0.406	0.355	11.938	15.7	300	47.1	0.128	4.37	53.5	48.5
	XS, 80S	0.500	0.438	11.750	19.2	362	56.7	0.160	4.33	65.4	47.0
	60	0.562	0.492	11.626	21.5	401	62.8	0.182	4.31	73.2	46.0
	80	0.625	0.547	11.500	23.8	439	68.8	0.204	4.29	80.9	45.0
		0.687	0.601	11.376	26.0	475	74.5	0.227	4.27	88.5	44.0
	100	0.750	0.655	11.250	28.3	511	80.2	0.250	4.25	96.2	43.0
		0.843	0.738	11.064	31.5	562	88.1	0.285	4.22	107	41.6
		0.875	0.766	11.000	32.6	579	90.8	0.298	4.21	111	41.1

TABLE A4 (Continued)
Properties and Weights of Pipe

Nominal Size Outside Diameter inches D	Weight Designation and/or Schedule Number	Average Wall Thickness inches t	Minimum Wall Thickness ($=\frac{7}{8}t$) inches t_m	Inside Diameter inches d	Cross-Sectional Metal Area sq. inches A	Moment of Inertia inches^4 I	Section Modulus inches^3 Z	Bend Characteristic per Unit Bend Radius 1/ft h/R	Radius of Gyration inches r_g	Weight of Pipe w_p lb per ft	Weight of Water w_w lb per ft
	120	1.000	0.875	10.750	36.9	642	101	0.348	4.17	125	39.3
	140	1.125	0.983	10.500	41.1	701	110	0.400	4.13	140	37.5
	160	1.312	1.149	10.126	47.1	781	123	0.481	4.07	160	34.9
	10	0.250	0.219	13.500	10.8	255	36.5	0.064	4.86	36.7	62.0
	20	0.312	0.273	13.375	13.4	315	45.0	0.080	4.84	45.7	60.6
	Std. 30	0.35	0.328	13.250	16.1	373	53.3	0.097	4.82	54.6	59.7
	40	0.438	0.382	13.125	18.7	429	61.4	0.114	4.80	63.4	58.6
	XS	0.500	0.438	13.000	21.2	484	69.1	0.132	4.78	72.1	57.5
	60	0.593	0.519	12.814	25.0	562	80.3	0.158	4.74	84.9	55.9
		0.625	0.547	12.750	26.3	589	84.1	0.168	4.73	89.3	55.3
14" 14.000	80	0.750	0.656	12.500	31.2	687	98.2	0.205	4.69	106	53.1
		0.875	0.766	12.250	36.1	781	112	0.244	4.65	123	51.1
	100	0.937	0.820	12.125	38.4	825	118	0.264	4.63	131	50.0
	120	1.093	0.956	11.814	44.3	930	133	0.315	4.58	151	47.5

232

Nom. size (OD)	Sched. No.	Desig.	t (col a)	(col b)	ID (col c)	(col d)	(col e)	(col f)	(col g)	(col h)	(col i)	(col j)
	140		1.250	1.094	11.500	50.1	1030	147	0.369	4.53	170	45.0
	160		1.406	1.230	11.188	55.6	1120	160	0.426	4.48	189	42.6
16″ 16.000	10		0.250	0.219	15.500	12.4	384	48.0	0.048	5.57	42.1	81.7
	20		0.312	0.273	15.376	15.4	474	59.3	0.061	5.55	52.3	80.5
	30	Std.	0.375	0.328	15.250	18.4	562	70.3	0.074	5.53	62.6	79.1
	40	XS	0.500	0.438	15.000	24.4	732	91.5	0.100	5.48	82.8	76.5
			0.625	0.547	14.750	30.2	894	112	0.127	5.44	103	74.1
	60		0.656	0.574	14.688	31.6	933	117	0.134	5.43	108	73.4
			0.750	0.655	14.500	35.9	1050	131	0.155	5.40	122	71.5
	80		0.843	0.738	14.314	40.1	1160	145	0.176	5.37	136	69.7
			0.875	0.766	14.250	41.6	1190	149	0.184	5.36	141	69.1
	100		1.031	0.902	13.938	48.5	1370	171	0.221	5.29	165	66.1
	120		1.218	1.066	13.564	56.6	1560	195	0.268	5.23	192	62.6
	140		1.438	1.258	13.124	65.8	1760	220	0.325	5.17	224	58.6
	160		1.593	1.394	12.814	72.1	1890	237	0.368	5.12	245	55.9
18″ 18.000	10		0.250	0.219	17.500	13.9	549	61.0	0.038	6.28	47.4	104
	20		0.312	0.273	17.376	17.3	679	75.5	0.048	6.25	59.0	103
		Std.	0.375	0.328	17.250	20.8	807	89.6	0.058	6.23	70.6	101
	30		0.438	0.382	17.124	24.2	932	104	0.068	6.21	82.2	99.7
		XS	0.500	0.438	17.000	27.5	1050	117	0.078	6.19	93.5	98.3
	40		0.562	0.492	16.876	30.8	1170	130	0.089	6.17	105	96.9
			0.625	0.547	16.750	34.1	1290	143	0.099	6.15	116	95.4
	60		0.750	0.656	16.500	40.6	1520	168	0.121	6.10	138	92.6
			0.875	0.766	16.250	47.1	1730	192	0.143	6.06	160	89.9
	80		0.937	0.820	16.126	50.2	1830	204	0.155	6.04	171	88.5
	100		1.156	1.012	15.688	61.2	2180	242	0.196	5.97	208	83.7
	120		1.375	1.203	15.250	71.8	2500	278	0.239	5.90	244	79.1
	140		1.562	1.367	14.876	80.7	2750	305	0.278	5.84	274	75.3
	160		1.781	1.558	14.438	90.8	3020	336	0.325	5.77	309	70.9

TABLE A4 (Continued)
Properties and Weights of Pipe

Nominal Size Outside Diameter inches D	Weight Designation and/or Schedule Number	Average Wall Thickness inches t	Minimum Wall Thickness $(=\frac{7}{8}t)$ inches t_m	Inside Diameter inches d	Cross-Sectional Metal Area sq. inches A	Moment of Inertia inches⁴ I	Section Modulus inches³ Z	Bend Characteristic per Unit Bend Radius 1/ft h/R	Radius of Gyration inches r_g	Weight of Pipe w_p lb per ft	Weight of Water w_w lb per ft
	10	0.250	0.219	19.500	15.5	757	75.7	0.031	6.98	52.7	129
	Std. 20	0.375	0.328	19.250	23.1	1110	111	0.047	6.94	78.6	126
	XS 30	0.500	0.438	19.000	30.6	1460	146	0.063	6.90	104	123
	40	0.593	0.519	18.814	36.2	1700	170	0.076	6.86	123	120
		0.625	0.547	18.750	38.0	1790	179	0.080	6.85	129	120
		0.750	0.655	18.500	45.4	2100	210	0.097	6.81	154	117
20"	60	0.812	0.711	18.376	48.9	2260	226	0.106	6.79	166	115
20.000	80	0.875	0.766	18.250	52.6	2410	241	0.115	6.77	179	113
		1.031	0.902	17.938	61.4	2770	277	0.138	6.72	209	109
	100	1.281	1.121	17.438	75.3	3320	332	0.175	6.63	256	103
	120	1.500	1.313	17.000	87.2	3760	376	0.210	6.56	296	98.3

		140	1.750	1.531	16.500	100	4220	422	0.252	6.48	341	92.6
		160	1.968	1.722	16.064	112	4590	459	0.291	6.41	379	87.8
24"	24.000	10	0.250	0.219	23.500	18.7	1320	110	0.021	8.40	63.4	188
		20 / Std.	0.375	0.328	23.250	27.8	1940	162	0.032	8.35	94.6	184
		XS	0.500	0.438	23.000	36.9	2550	213	0.043	8.31	125	180
		30	0.562	0.492	22.875	41.4	2840	237	0.049	8.29	141	178
			0.625	0.547	22.750	45.9	3140	261	0.055	8.27	156	176
		40	0.687	0.601	22.625	50.3	3420	285	0.061	8.25	171	174
			0.750	0.655	22.500	54.8	3710	309	0.067	8.22	186	172
		60	0.968	0.847	22.064	70.0	4650	388	0.088	8.15	238	166
		80	1.218	1.066	21.564	87.2	5670	473	0.113	8.07	296	158
		100	1.531	1.340	20.938	108	6850	571	0.146	7.96	367	149
		120	1.812	1.586	20.376	126	7820	652	0.177	7.87	429	141
		140	2.062	1.804	19.876	142	8630	719	0.206	7.79	483	134
		160	2.343	2.050	19.314	159	9460	788	0.240	7.70	542	127
30"	30.000	10	0.312	0.273	29.376	29.1	3210	214	0.017	10.5	98.9	294
			0.375	0.328	29.250	34.9	3830	255	0.021	10.5	119	291
		20	0.500	0.438	29.000	46.3	5040	336	0.028	10.4	157	286
			0.562	0.492	28.875	52.0	5640	376	0.031	10.4	177	284
		30	0.625	0.547	28.750	57.6	6220	415	0.035	10.4	196	281
			0.750	0.655	28.500	68.9	7380	492	0.042	10.3	234	277

Reproduced from M. W. Kellogg Co. *Design of Piping Systems*, Wiley, New York,

**TABLE A5   Sample Calculations for
Branch Reinforcement***

The following examples are intended to illustrate the application of the rules
and definitions in B31.3 Sec. 304.3.3 (Eqs. 2.10 through 2.13) for welded
branch connections. (No metric equivalents are given.)

Example 1

An NPS 8 run (header) in an oil piping system has an NPS 4 branch at right
angles (see Fig. A1). Both pipes are schedule 40 API 5L Grade A seamless.
The design conditions are 300 psig at 400°F. The fillet welds at the crotch are
minimum size. A corrosion allowance of 0.10 in. is specified. Is additional
reinforcement necessary?

Solution:
From Appendix A Table 1, of ANSI/ASME B31.3 (Appendix A3) $SE =$
16.0 ksi.

$$T_h = (0.322)(0.875) = 0.282 \text{ in.}$$

$$T_b = (0.237)(0.875) = 0.207 \text{ in.}$$

$$L_4 = 2.5(0.282 - 0.1) = 0.455 \text{ in., or}$$
$$\qquad 2.5(0.207 - 0.1) + 0 = 0.268 \text{ in., whichever is less}$$

$$L_4 = 0.268 \text{ in.}$$

$$d_1 = [4.5 - 2(0.207 - 0.1)]/\sin(90°) = 4.286 \text{ in.}$$

$$d_2 = (0.207 - 0.1) + (0.282 - 0.1) + \frac{4.286}{2} = 2.432 \text{ in.}$$
$$\qquad \text{or } d_1, \text{ whichever is greater}$$

$$d_2 = 4.286 \text{ in.}$$

$$t_h = \frac{(300)(8.625)}{(2)(16,000) + (2)(0.4)(300)} = 0.080 \text{ in.} \tag{2.1}$$

$$t_b = \frac{(300)(4.500)}{(2)(16000) + (2)(0.4)(300)} = 0.042 \text{ in.}$$

For fillet weld $t_c = 0.7(T_b)$

$$t_c = 0.7(0.237) = 0.166 \text{ in., or}$$

$$= 0.25, \text{ whichever is less}$$

$$t_c = 0.166 \text{ in.}$$

*Appendix H of B31.3 Code.

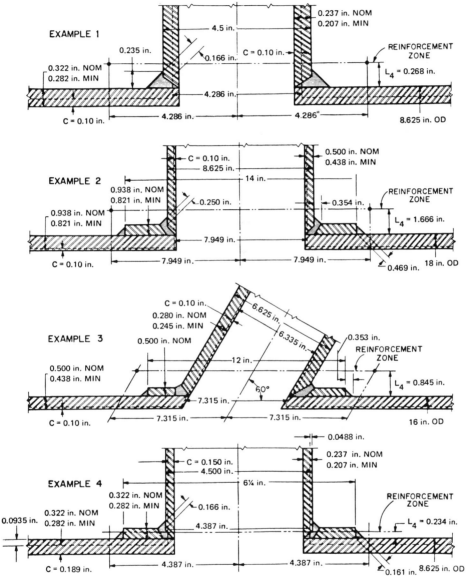

FIGURE A1 Illustration for branch reinforcement.

Minimum leg dimension of fillet weld $= \dfrac{0.166}{0.707} = 0.235$ in.

Thus, the required area, $A_1 = (0.080)(4.286)[2 - \sin(90°)] = 0.343$ sq. in.

The reinforcement area

in run wall, $A_2 = (4.286)(0.282 - 0.08 - 0.10) = 0.437$ sq. in.
in branch wall, $A_3 = (2)(0.268)[(0.207 - 0.042) - 0.10]$ $= 0.035$ sq. in.
in branch welds, $A_4 = (2)(\frac{1}{2})(0.235)^2$ $= 0.055$ aq. in.
The total reinforcement area $= 0.527$ sq. in.

This is more than 0.343 sq. in. so that no additional reinforcement is required to sustain the internal pressure.

Example 2

There is an NPS 8 branch at right angles to an NPS 18 header (Fig. A1). Both run and branch are of Schedule 80 ASTM A335 Grade P11 seamless pipe. The connection is reinforced by a ring 14 in. O.D. (measured along the run) cut from a piece of NPS 18 Schedule 80 ASTM A335 Grade P12 seamless pipe and opened slightly to fit over the run pipe. The fillet welds have the minimum dimensions. A corrosion allowance of 0.10 in. is specified. What is the maximum permissible normal operating pressure if the normal operating temperature is 1005°F?

Solution:
From Appendix A Table 1 of B31.3 (Appendix A3) $SE = 7.68$ ksi for Grade P11 and $SE = 7.38$ ksi for Grade P12, both at 1005°F.

Leg Dimensions of Welds: (See Fig. A1 for weld dimensions)
$\dfrac{t_c}{0.707} = \dfrac{0.250}{0.707} = 0.354$ in. (Due to $\frac{1}{4}$ in. minimum for weld)

$\dfrac{(0.5)(0.938)}{0.707} = 0.663$ in.

$T_h = (0.938)(0.875) = 0.821$ in. (Due to M.T of 12.5%)

$T_b = (0.500)(0.875) = 0.438$ in.

$t_r = (0.938)(0.875) = 0.821$ in.

$L_4 = 2.5(0.438 - 0.10) + 0.821 = 1.666$ in.
       [This is smaller than $2.5(0.821 - 0.10) = 1.80$ in.]

$d_2 = d_1 = 8.625 - 2(0.438 - 0.10) = 7.949$ in.
y coefficient for 1005°F $= 0.7$ (Table 2.1)
$SE = 7.68$ ksi (Appendix A3)

$$t_h = \frac{18P}{(2)(7680) + (2)(0.7)(P)} \tag{2.1}$$

$$t_b = \frac{(8.625)(P)}{(2)(7680) + (2)(0.7)(P)}$$

Using the symbol $q = \dfrac{P}{15,360 + 1.4P}$, we can briefly write

$t_h = 18q$ and $t_b = 8.625q$

The required area, $A_1 = 7.949\ t_h = 143.08q$. $\hspace{2em}$ (2.10)
The reinforcement area
in run wall, $A_2 \hspace{1.5em} = (7.949)(0.821 - 18q - 0.10) = 5.731 - 143.08q$

$$\tag{2.12}$$

in branch wall, $A_3 = (2)(1.666)(0.438 - 8.625q - 0.10) = 1.126 - 28.73q$

$$\tag{2.13}$$

in ring, $A_4 \hspace{2em} = (0.821)(14 - 8.625)(7250/7570) \hspace{1em} = 4.226$
in fillet welds, $A_4 = (2)(\tfrac{1}{2})(0.354)^2 + (2)(\tfrac{1}{2})(0.663)^2 \hspace{1em} = 0.565$
The total reinforcement area $\hspace{8em} = 11.648 - 171.818q$

At the maximum permissible normal operating pressure, the required area
and the reinforcement area are equal; thus:

$$143.08q = 11.648 - 171.818q;\ 314.898q = 11.648;\ q = 0.0370$$

But also,

$$q = \frac{P}{15,360 + 1.4P}$$

Thus, $P = (0.0370)(15360 + 1.4P) = 568.32 + 0.0518P$

$0.948P = 568.32$

$\hspace{2em} P = 699.494$ psig which is the maximum permissible normal operating
$\hspace{3em}$ pressure.

Example 3

An NPS 6 Schedule 40 branch has its axis at a 60 deg. angle to the axis of an
NPS 16 Schedule 40 run (header) in an oil piping system (Fig. A1). Both
pipes are API 5L Grade A seamless. The connection is reinforced with a ring
12 in. O.D. (measured along the run) made from $\tfrac{1}{2}$ in. ASTM A285 Grade C
plate. All fillet welds are equivalent to 45 deg. fillet welds with $\tfrac{3}{8}$ in. legs.

Corrosion allowance $= 0.10$ in. The design pressure is 500 psig at 700°F. Is the design adequate for the internal pressure?

Solution:

The allowable stress values from Appendix A Table B31.3 (Appendix A3) are: for pipe, $SE = 14.4$ ksi; for ring, $SE = 14.4$ ksi.

$$T_h = (0.500)(0.875) = 0.438 \text{ in.}$$

$$T_b = (0.280)(0.875) = 0.245 \text{ in.}$$

$$t_r = 0.500 \text{ in.}$$

$$L_4 = 2.5(0.0245 - 0.10) + 0.500$$

$$= 0.8625. \text{ This is greater than } 2.5(0.438 - 0.10) = 0.845 \text{ in.}$$

$$t_h = \frac{(500)(16)}{(2)(14{,}400) + (2)(0.4)(500)} = 0.274 \text{ in.}$$

$$t_b = \frac{(500)(6.625)}{(2)(14{,}400) + (2)(0.4)(500)} = 0.113 \text{ in.}$$

$$d_2 = d_1 = \frac{6.625 - 2(0.245 - 0.10)}{\sin 60°} = \frac{6.335}{0.866} = 7.315 \text{ in.}$$

The required area, $A_1 = (0.274)(7.315)(2 - 0.866) = 2.27$ sq in. $\qquad$ (2.10)

The reinforcement area

in run wall, $A_2 \quad = (7.315)(0.438 - 0.274 - 0.10) \quad = 0.468$ sq. in.

$$\text{(2.12)}$$

in branch wall, $A_3 \quad = (2)\dfrac{0.845}{0.866}(0.245 - 0.113 - 0.10) \quad = 0.062$ sq. in.

$$\text{(2.13)}$$

in ring, $A_4 \qquad = 0.500\left(12 - \dfrac{6.625}{0.866}\right) \qquad = 2.175$ sq. in.

in fillet welds, $A_4 = (4)(\frac{1}{2})(\frac{3}{8})^2 \qquad = 0.281$ sq. in.
Total reinforcement area $\qquad = 2.986$ sq in.

This total is greater than 2.27 sq. in., so that no additional reinforcement is required.

Example 4

An NPS 8 run (header) in an oil piping system has an NPS 4 branch at right angles (Fig. A1). Both pipes are Schedule 40 API 5L Grade A seamless. The design conditions are 350 psig at 400°F. It is assumed that the piping system is to remain in service until all metal thickness, in both branch and header, in excess of that required by Equation 2.1 has corroded away. What reinforcing is required for this connection?

Solution:
The allowable stress value from Appendix A, Table 1 of B31.3 (Appendix A3) is $SE = 16.0$ ksi.

$$t_h = \frac{(350)(8.625)}{(2)(16,000) + (2)(0.4)(350)} = 0.0935 \text{ in.}$$

$$t_b = \frac{(350)(4.500)}{(2)(16,000) + (2)(0.4)(350)} = 0.0488 \text{ in.}$$

$$d_1 = 4.500 - (2)(0.0488) = 4.402 \text{ in.}$$

Required reinforcing area, $A_1 = (0.0935)(4.402) = 0.412$ sq. in.

Try fillet welds only.

$$L_4 = (2.5)(0.0935) = 0.234 \text{ in.}$$
$$\text{or}$$
$$(2.5)(0.0488) = 0.122 \text{ in.} \quad \text{use } 0.122 \text{ in.}$$

Due to limitation in the height at the reinforcing zone, no practical fillet weld size will supply enough reinforcement area; therefore, the connection must be reinforced by a ring. Try a ring of $6\frac{1}{4}$ in. O.D. (measured along the run). Assume the ring to be cut from a piece of NPS 8 Schedule 40 API 5L Grade A seamless pipe and welded to the connection with minimum size fillet welds.

Min. pad thickness, $t_r = (0.322)(0.875) = 0.282$ in.

New $L_4 = (2.5)(0.0488) + 0.282 = 0.404$ in.
$$\text{or}$$
$$(2.5)(0.0935) \qquad = 0.234 \text{ in.} \quad \text{use } 0.234 \text{ in.}$$

Reinforcement area in the ring (considering only the thickness within L_4):

$$X_1 = 0.234(6.25 - 4.5) = 0.410 \text{ sq. in.}$$

Leg Dimension of Weld:

$$\frac{(0.5)(0.322)}{0.707} = 0.228 \text{ in.}$$

Reinforcement area in fillet welds:

$$X_2 = (2)(\tfrac{1}{2})(0.228)^2 = 0.052 \text{ sq. in.}$$

Total Reinforcement Area, $A_4 = X_1 + X_2 = 0.462$ sq. in.

This total reinforcement area is greater than the required reinforcing area; therefore a reinforcing ring of $6\tfrac{1}{4}$ in. O.D., cut from a piece of NPS 8 Schedule 40 API 5L Grade A seamless pipe and welded to the connection with minimum size fillet welds would provide adequate reinforcing for this connection.

Example 5 (Not illustrated)

An NPS $1\tfrac{1}{2}$ 3000 lb forged steel socket welding coupling has been welded at right angles to an NPS 8 Schedule 40 header in oil service. The header is ASTM A53 Grade B seamless pipe. The design pressure is 400 psi and the design temperature is 450°F. The corrosion allowance is 0.10 in. Is additional reinforcement required?

Solution:
No. Since branch is less than NPS 2 (according to B31.3 Section 304.3.2(b)) the design is adequate to sustain the internal pressure and no calculations are necessary. It is presumed, of course, that calculations have shown the run pipe to be satisfactory for the service conditions according to Equations 2.1, 2.3 and 2.4.

INDEX

243